CARS

CARS

ACCELERATING THE MODERN WORLD

EDITED BY
BRENDAN CORMIER AND LIZZIE BISLEY
ASSISTED BY ESME HAWES

V&A PUBLISHING

Published to accompany the exhibition
Cars: Accelerating the Modern World at
the Victoria and Albert Museum, London,
from 23 November 2019 to 19 April 2020.

Supported by:

BOSCH
Invented for life

First published by V&A Publishing, 2019
Victoria and Albert Museum
South Kensington
London SW7 2RL
www.vandapublishing.com

Distributed in North America by Abrams,
an imprint of ABRAMS

ISBN 9781–85177–967–3

10   9   8   7   6   5   4   3   2   1
2023   2022   2021   2020   2019

Designer, including cover design: Jonathan Abbott
Copy-editor: Mandy Greenfield
Origination: XY Digital
Index: Hilary Bird
New photography by Paul Robins and
Pip Barnard, V&A Photographic Studio

Printed by Everbest, China

MIX
Paper from
responsible sources
FSC® C124385

V&A Publishing
Supporting the world's leading
museum of art and design,
the Victoria and Albert
Museum, London

←← **page 2** Auto 'Death
Dodgers' at the 1939
New York World's
Fair, by an unknown
photographer, c.1939

→ **page 5** Jaguar E-Type
on the M1, photograph
by Brian Duffy, 1961

→→ **page 6** Handsworth
Riots, Birmingham,
photographs by
Pogus Ceasar, 1985

↑ *110 Junction*, photograph by Matthew Porter, 2010

Director's Foreword
Tristram Hunt

The invention of the car, over 100 years ago, has had a profound impact on the history of modern design. From its earliest beginnings, the car swiftly reshaped landscapes, cities, fashion, film, and even whole economies. Today it is almost impossible to imagine a world without the car.

Since its inception, the car has been an object that sparks excitement and wonder. It can represent independence and freedom, opening up a whole new world of possibilities to the curious traveller, and can similarly serve as the pinnacle of luxurious design. But in today's global 'car culture', questions also abound concerning the future of our transport systems. With the threat of climate change looming, there is an urgent need for systems of mobility that are radically more sustainable and efficient. Thoughts of electric cars, self-driving cars and even flying cars are beginning to feel more like reality than science fiction, and ever-more exciting ideas are entering the public consciousness. From the more prosaic suggestions (such as bike and e-scooter sharing systems), to wilder leaps of the imagination (such as Silicon-Valley dreams of tunnels and tubes criss-crossing the planet), we are only just beginning to consider what travel and transport might look like in the next 100 years.

Design and innovation are so often followed by unintended consequences, reshaping the world in ways their creators could never have imagined. The car has had an unprecedented influence not only on our transport systems, but in other design spheres such as advertising, furniture and photography – as this project will show. But as we rethink our transport systems in the twenty-first century, we must also be mindful of the mistakes of a past century of car-centric development. With the exhibition *Cars: Accelerating the Modern World,* we are not only celebrating the ingenuity of the car and its fascinating impact on the world, but also confronting some of the more undesirable outcomes brought about by the invention of the automobile.

This exhibition and catalogue mark the first time that the V&A has focused on the automobile as a design object. There has been a perceived tendency in the past for cultural institutions to view science and technology as distinctly separate entities from the arts, but – going right back to our roots in the 1851 Great Exhibition – it has always been the V&A's mission to explore design in all its forms, showcasing the most innovative and exciting developments of the day, and interrogating them for a curious public. As an object that so perfectly straddles the spheres of both art and technology, with its own complex history, we are happy to welcome the car into the museum. And that might well be its future.

The V&A is delighted to be working with Bosch as the supporting partner of *Cars: Accelerating the Modern World.* Bosch is a particularly fitting partner for this exhibition, as their development of electrical safety features drastically improved early motoring safety, which in turn led to significant developments in automated driving. We are grateful for their generous support.

Introduction

Brendan Cormier and Lizzie Bisley

Well we know where we're going
But we don't know where we've been
And we know what we're knowing
But we can't say what we've seen
And we're not little children
And we know what we want
And the future is certain
Give us time to work it out
'Road to Nowhere', Talking Heads, 1985

The car has always been a dream. Since the late nineteenth century, when the first automobiles began to emerge, people have looked to cars as signs of the future: portents for a new world of technological utopia. Often embodying an image of constant, fast, free-flowing movement, the car has symbolized all kinds of social, political, personal and economic transformation. It has stood for the possibility of a new way of living, while also being an active agent in shaping the systems, structures and images that have defined the modern world.

As we come to the end of the 2010s and the car approaches its 150th anniversary, the short but significant history of this most highly keyed machine appears to be entering a period of drastic change. With oil stocks running out, the climate warming and massive crowding and congestion on our roads and in our cities, designers, manufacturers, town planners, politicians and economists are all aware that the design of the car needs to shift, if we want it to survive at all. The vehicle of the future, increasingly seen as electric powered and autonomously driven, is being actively developed and experimented with by manufacturers and designers around the world.

What is striking, however, is the extent to which today's 'radical' new designs recycle and repeat those visions of the past 150 years. Elon Musk's 'Boring Project', a scheme to move autonomous cars at breakneck speed along dedicated tunnels, bears uncanny similarities to General Motors' Firebird cars of the 1950s and their accompanying plan for an autonomous system of high-speed highways. Debate around a contemporary move towards electric-powered cars provides a mirror image of the same discussions held in the 1880s and '90s around the (then-dominant) electric automobile. And the flying car, so long the fantasy of science-fiction writers around the planet, is yet again being proffered as the ultimate dream-solution to overcrowding and over-whelming traffic – with dozens of new start-ups racing to secure angel investment for what they promise will be the first viable model.

At this crucial point then, when cars will need to change more drastically and more quickly than they have since their early years, it is important that we fully understand where we are coming from – the brief but dense history of how the car has been imagined, made and sold, and how the world changed in response – in order to better envision what may come next. As the art historian Erwin Panofsky so elegantly put it, 'We often invent the future out of fragments from the past.'[1] This book, as well as the exhibition that it accompanies, is an attempt to collect those fragments: a sampling of stories and historical cases that might help us determine where we want to go from here.

In approaching this dauntingly vast subject, we have focused our attention on the car's agency as a piece of design. It is not an exaggeration to say that there is no other object that has had such an immense impact on the past century. Through expected and unexpected ripples, and in ways that have been both highly anarchic and very densely planned, the car has determined much of the world in which we now live. This extends from our experience of speed to the ways in which we make and buy products, and the very landscapes that we inhabit. At a moment of intense over-production and over-consumption, when many designers and writers are grappling with the question of what design can offer the planet (other than a profusion of stuff), the car offers a unique case study for the power of objects and design decisions to change the course of the world.

By framing the car in relation to its wider history, we have tried to reintroduce it into the world of design. Designers, design historians and, in particular, design museums have long had an uncomfortable relationship with cars, seeing them largely as either the purview of technical, scientific histories or as purely fetishistic objects of beauty and styling. Car exhibitions, when held in design museums, have tended to focus entirely on form, or occasionally on individual (messianic) designers

and marques. In this book we seek to reposition the car in design historical discourse, claiming a place for it as a serious object of study: one that can offer important, fresh perspectives on the social, political, economic and environmental histories of the twentieth and early twenty-first centuries.

The book begins with a visual essay of depictions of the future, showing how a collective dream of fluidity and frictionless movement has been embodied in the image of the car for over a century; we then address five key areas in which cars have driven a systematic acceleration and deep-rooted change. The book's five main chapters look at: the experience of speed; production and manufacturing; consumption and styling; the environments and landscapes of the car; and the construction of the nation.

Accompanying each of these chapters is a short piece that addresses the theme from a contemporary perspective. While the main chapters have been written by the exhibition curators, we've invited critics, journalists and design writers to write these pieces, which grapple with the major questions being raised by changes in the car industy today.

The contemporary texts address big subjects – such as the future of cars in the city, or the environmental impact of the lithium-ion battery – while at the same time adding personal voices and geographic scope. The car is one of the most truly global objects of the twentieth and twenty-first centuries: driven, adapted and dreamed of in every part of the world. While this project focuses primarily on the car's North American and European history, we have tried to flesh this out with glimpses into important, and highly specific, histories of the car in other places.

The car is a huge, messy, exhilarating, disturbing and deeply personal affair. Although we cannot hope this book will tell all of the important stories that spin out from this one object, what we have tried to do is claim a place for cars in the network of design and consequence that is etched on to the face of the modern globe. At once a cautionary tale and a joyful, uncanny sweep, the story of the car cries out for urgent examination by anyone with an interest or a stake in the wider world of contemporary design.

GOING FAST

FAST FUTURES

Brendan Cormier

Throughout the twentieth century the car figured prominently in the imagery of science fiction. To look through these images is to understand how we have collectively fantasized about the future of mobility for more than 100 years. A common motif across all these images is that technology – primarily in the form of a car – will bring us a new way of getting around: one that is fast, frictionless and free, without a traffic jam in sight. From steampunk fantasies to Afrofuturism, Hollywood blockbusters to concept-car extravaganzas, speed has figured at the heart of our mobile dreams.

↑ 1 Testing the Firebird XP-21 on the racetrack in Phoenix, Arizona, 1954

Inspired by Flight

Automotive designer Harley Earl is seen here in 1954 with racing driver Mauri Rose, testing General Motors' concept car, the Firebird XP-21 on the track in Phoenix, Arizona. The Firebird series of concept cars drew heavily from the look and technology of jet fighters from the 1940s and '50s.

Postcards from Future Moscow

In 1914 various artists were commissioned for a series of postcards to imagine what Moscow might look like in the twenty-third century AD. The postcards portray several futuristic and anachronistic elements together vying for space in a still bustling and chaotic city.

↙ 2 'The Central Railway Station', from the series *Moscow of the Future*, 1914

↓ 3 'The Red Square', from the series *Moscow of the Future*, 1914

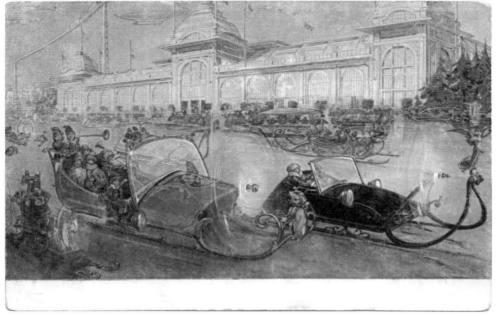

Dreaming of the Motorway

John Douglas-Scott-Montagu, 2nd Baron Montagu of Beaulieu, was an early advocate of motoring in the UK. In the 1920s he also championed the idea of dedicated car-only motorways for the country. These illustrations by Ferdinand Fermo Fissi, who collaborated regularly with Montagu, depict such a landscape, where highways straddle the built fabric of the city.

Landscapes of Streamlined Motion

Alexander Leydenfrost cut his teeth as an industrial designer working on, among other things, the streamlined redesign for the Chrysler Airflow. In 1939 he became a full-time illustrator, drawing from his previous career to render landscapes of fluid and streamlined motion for several science fiction and popular science magazines.

→ **6** *City of the Future (Rush Hour)*, by Alexander Leydenfrost, c.1949

Illustrations in Japanese Children's Books

From the 1950s to the 1970s in Japan, many children's books were published that speculated on what the future twenty-first century might look like. This illustration, taken from the book *Tanoshii Yonensei* (Happy 4th Year Student), condenses many classic science fiction elements into one image, including hover cars and smooth curvilinear transport infrastructure.

Popular Science

Established in 1872, *Popular Science* magazine was an important American quarterly for the dissemination of scientific knowledge to a mainstream audience. In the twentieth century it changed direction and became a celebrated outlet for the expression of more radical and imaginative designs, especially futuristic car concepts.

↑ **7** 'Tokyo of 2061', by Tenan Ito, published in *Tanoshii Yonensei* (Happy 4th Year Student), 1961

→ **8** Cover illustration, by Edgar Franklin Wittmack, for *Popular Science*, November 1933

SEE PAGE 47

Solving Road Rage

On 16 December 1962, the Italian weekly newspaper *La Domenica Del Corriere* published these illustrations by Walter Molino. The front cover depicts a nightmarish traffic jam, while the back cover presents a futuristic solution, in which commuters travel seamlessly and unperturbed in self-contained pods.

↓→ **9** 'Transport of the Future' (front and back cover), by Walter Molino, published in *La Domenica Del Corriere*, 16 December 1962

La tragedia del Nova Scotia
Terrificante testimonianza alle pagg. 12-13

DOMENICA DEL CORRIERE

Una favola più grande di lui
Articolo di Dino Buzzati a pagina 5

In città gireremo così?

Ecco come potrebbe essere alleggerito, se non del tutto risolto, il problema del traffico nelle città: anzichè le attuali ingombranti vetture, delle minuscole auto monoposto che occupano una minima superficie e che potrebbero essere battezzate « singolette ». Walter Molino ha immaginato qui l'aspetto della stessa strada della prima tavola qualora venisse adottata su larga scala la nuova soluzione. Serv. alle pagg. 6-7.

Soviet Visions of Progress

Tekhnika Molodezhi (Technology Youth) was a Soviet popular science magazine, founded in 1933. Throughout its several decades of publication, it took a particular interest in featuring work that reimagines mobility and how we might move around in the future, offering a Soviet counterpoint to western sci-fi images.

↑ **10** Cover illustration, by Konstantin Artseulov, for *Tekhnika Molodezhi*, no. 2, 1949

→ **11** 'Moscow of the third millennium', by Nikolay Nedbaylo (1971), published in *Tekhnika Molodezhi*, no. 6, 1972

An Encyclopaedia of Transport

The *Encyclopédie des transports présents et à venir* was a spin-off from the popular graphic novel series *Les Cités obscures* by Belgian comic artist François Schuiten, created in collaboration with writer Axel Wappendorf. It depicts uncanny and surreal transport devices that reimagine and combine nineteenth- and early twentieth-century technologies in novel ways.

← **12** *'Le Vélocipède Alaxien'*, by François Schuiten, published in *Encyclopédie des transports présents et à venir*, 1988

↙ **13** *'Le Tripode Aquatique'*, by François Schuiten, published in *Encyclopédie des transports présents et à venir*, 1988

↓ **14** *'Le Torpillard'*, by François Schuiten, published in *Encyclopédie des transports présents et à venir*, 1988

New Makoko Village
[LAGOS 2081 A.D.]

Our Africa 2081 A.D.

In 2013 Nigerian-born fashion designer Walé Oyédijé commissioned the artist Olalekan Jeyifous to illustrate a lookbook for his Ikiré Jones label. Titled 'Our Africa 2081 A.D.', it features men in Ikiré Jones clothing, standing in lively urban settings that combine both utopian and dystopian visions of what a future African city might look like. Here, the floating squatter settlement of Makoko in Lagos is juxtaposed with towering communications infrastructure and flying vehicles in the background.

At the Movies

Hollywood has often used novel transport to quickly establish a futuristic city setting. In Luc Besson's *The Fifth Element* (1997), the director amplifies the filmic trope of the car chase by introducing flying cars that can weave and twist in multiple directions through a dense urban environment.

THE DESIGN OF SPEED

Brendan Cormier

↑ **17** Belgian race–car driver Camille Jenatzy sitting in *La Jamais Contente*, the first car to reach 100 km/h in 1899

On 29 April 1899 the young Belgian entrepreneur Camille Jenatzy did something that had never been done before. On a public road outside Achères – a town on the outskirts of Paris – Jenatzy propelled his body 105.882 kilometres per hour. He did so using just a metal tube, a chassis with four rubber wheels and an electric engine, thus exceeding the 100 km/h mark for the first time, and breaking the world land-speed record [17]. The vehicle that won him the honour was called *La Jamais Contente*, which translates as 'the never satisfied'. Although it was jokingly a reference to his wife,[1] the name also speaks of Jenatzy's stubborn drive to improve on his own speed records (he had already set two in the previous four months). But 'never satisfied' was also prophetic, as it would go on to perfectly symbolize an entire history of professional motor racing – the compulsion to go ever faster, which has afflicted the hearts and minds of thousands of engineers, designers, mechanics and drivers over the past 120 years. This lust for speed has had a transformative effect – not only has it been instrumental in defining the design and technology of automobiles for more than a century, but also, in so doing, it has fundamentally changed the way we think about and experience the world.

Speed is All Around

On 29 January 1886, just 13 years prior to Jenatzy's legendary race, the Mannheim engineer Karl Benz filed a patent for a 'gas-powered vehicle' with reference number DRP 37435. The patent described what would widely be regarded as the first-ever production automobile – the Benz Patent Motorwagen. The three-wheeled carriage with a rear-mounted engine contained novel innovations, including a steel tube construction, steel-spoked wheels with solid rubber tyres and tiller steering [18].[2] Blazing speed, however, was not its strong point. When Benz unveiled the car to the wider public on the Ringstrasse in Mannheim on 3 July 1886, its single-cylinder combustion engine could barely muster 16 km/h.[3] Early motoring was off to a leisurely pace.

In fact most early motoring was typified by modest-to-slow speeds. In the UK, the first person to be convicted of a speeding violation – Walter Arnold of East Peckham – was caught on 28 January 1896 travelling at an egregious 13 km/h.[4] Later that same year an Irish woman named Bridget Driscoll had the unfortunate fate of becoming the first pedestrian in the country ever to be killed by a moving car. As she strolled with her teenage daughter and friend on the grounds of the Crystal Palace in London, a car giving demonstration rides struck her down at what one witness described as 'a reckless pace'. The driver claimed to have been driving at 6.4 km/h, while the maximum speed it could possibly go was 13 km/h.[5] Even *La Jamais Contente* three years later, hurtling Camille Jenatzy forward at impressive new automotive speeds, did not, in a way, really set a new land-speed record. Trains had broken the 100 km/h mark nearly 50 years earlier.

↓ 18 Bertha Benz with her husband, the inventor Karl Benz, driving a Benz Viktoria in 1894

But if cars were initially slow at the end of the nineteenth century, the rest of the world was moving incredibly fast. And it was from this context of acceleration that early automotive engineers and inventors took inspiration. All around, new innovations were helping to collapse notions of space and time, in ways never previously imagined. Steam power, for instance, and its use in train technology, indelibly altered the geography of the nineteenth century. As rail networks expanded across the world, and as train technology enabled faster rides, journeys that might have taken weeks were shortened to days and hours. In a series of illustrations the French satirist Honoré Daumier lampooned the state of train travel in France in the 1840s, which by then had become commonplace. In one image he portrayed the horror and comedy of the experience of out-of-control speed, depicting errant top hats, flailing limbs and well-dressed men being flung off an open-topped train carriage in what we are to assume was a typical daily French commute [19].

The way that news and messages were transmitted from one place to another was also undergoing a radical shift. By the 1830s the first electric telegraphy networks were being laid out, introducing a novel method of instantaneous communication. By the end of the nineteenth century Guglielmo Marconi was pioneering his first model radios, while Alexander Graham Bell was patenting the first practical telephone. Communication infrastructure proliferated across the world, in vast coordinated networks, so that by the end of the century information could travel at speeds and across distances as never before.

Mora 707 BROADWAY N. Y.

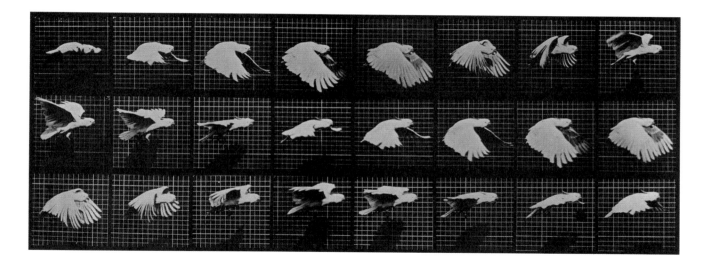

← **20** Alice Vanderbilt as 'Electric Light' at the Vanderbilt Ball, dress designed by Charles Frederick Worth, 1883

↑ **21** 'Bird in Flight', from photographer Eadweard Muybridge's series *Animal Locomotion*, published in 1887, in which he used a photographic technique to capture and visualize fast-moving action for the first time

Likewise, electricity was having a dramatic effect on the pace of production and on the frenzy of city life. In 1882 the Edison Electric Light Station was opened in London, the world's first coal-fired public power station.[6] The idea was to create a localized source of electricity for local residences, but also to power 3,000 electric incandescent lamps, from Holborn Circus to St Martin's Le Grand. The business operation ran at heavy losses and the station was forced to close just four years later.[7] But the precedent had been set, and the demand for local power stations providing public utilities, factories and residences with a steady supply of electricity quickly spread to cities around the world. Electric street lights, and the illumination of interiors at night, meant that the bustle of city life could now extend to a full 24 hours. Factories could also operate round the clock, through shift work, increasing the flow of goods and labourers to and from the factory floor.

Enthusiasm for electricity was such that, in 1883, Alice Vanderbilt, matriarch of the Vanderbilt family – one of the wealthiest families in the world – decided to dress up as electricity itself [20]. She commissioned Charles Frederick Worth to design the 'Electric Light' dress, a gown made of gold and silver thread, which she would wear to attend a masquerade ball thrown by her sister-in-law. It came with a built-in battery and lamp, which she could raise to mimic the Statue of Liberty.[8]

Methods of visualizing speed were changing as well. Photography, invented in the first half of the nineteenth century, had made significant advances by the end of the century. So much so that experimental photographers like Eadweard Muybridge were now able to capture the image of speed in revelatory ways. His motion-study photographs, in which simple actions undertaken by animals and humans were broken down into a series of freeze-frames [21], revealed the intricate details of biomechanics, but also the limitations of human perception to process actions beyond a certain velocity.

In the final years of the century the science and spectacle of speed merged, through the popularization of the moving image. Auguste and Louis Lumière were early pioneers of the medium, holding small public screenings of their experiments in cities around the world in order to drum up enthusiasm for the medium. One such film, *L'Arrivée d'un train en gare de La Ciotat* (1896), is a 50-second clip featuring a train arriving at a station – an apocryphal recounting of the screening claims that audience members were so frightened by the image of an oncoming train that they ran out of the room.[9] An even more mesmerizing film of theirs featured the Serpentine Dance by Loie Fuller. Fuller was an American dancer whose characteristic frenzied choreography, enhanced by the whiplash motion of her silk costumes and multi-coloured shifting lighting, made her a mainstay of the Paris arts and performance scene [23]. The 1896 film of her entrancing dance perfectly encapsulates, in a way, the dizzying ambience and electrifying pace that characterized the close of the nineteenth century.

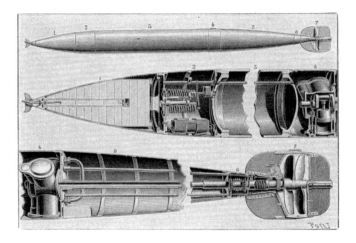

↑ 22 Illustration of a Whitehead torpedo, published in the French magazine *La Nature*, May 1891

→ 23 Portrait of Loie Fuller, capturing her frenetic and blurred choreography, which captivated Parisian audiences at the turn of the century

These innovations all helped to acclimatize a broader public to new experiences of speed, transforming the idea from something that inspired terror (recalling Daumier's image of the train) to something to be desired (Fuller's seductive Serpentine Dance).

Early car developments took inspiration from, and capitalized on, this context. Recall how *La Jamais Contente* – and, indeed, many early cars – ran on an electric engine. That was simply because electricity was the dominant technology that was driving much of the innovation of the era. (The petrol engine would remain a marginal technology until the dawn of the twentieth century.) *La Jamais Contente* also took its cue from another nascent nineteenth-century technology: the torpedo. The first modern self-propelled torpedo, the 'Whitehead torpedo', was invented by Robert Whitehead in 1866 [22].[10] By the 1880s navies around the world were using Whitehead torpedoes, which could now travel at speeds of over 30 km/h, to strike fear into the hearts of their opponents. The fact that the position of *La Jamais Contente*'s driver betrayed the actual aerodynamics of its torpedo-shaped body was beside the point: it was the image of the torpedo itself, as a stealth weapon of speed, that had caught the imagination of the designer.

Having been introduced in the nineteenth century to new experiences of speed in various new forms, the public had cultivated a strong appetite for it by the beginning of the twentieth century. The car thus arrived at the perfect moment, as an outlet for people to experience speed on their own terms. So while the car did not invent speed, it did help to democratize it, promising to put it in the hands of individuals – and, once behind the driver's wheel, the first logical thing to do was to race.

The Spectacle of the Race

La Jamais Contente's triumphant 100 km/h feat was not Jenatzy's first race. For the past year he had been engaged in a fierce rivalry with the wealthy aristocrat Gaston de Chasseloup-Laubat, who had been competing primarily with a Jeantaud electric vehicle. The initial idea to set up speed trials came from the magazine *La France Automobile*, and so in December 1898 the first official land-speed record was recorded, with the honour going to de Chasseloup-Laubat, who managed to average a speed of 63.13 km/h. Jenatzy and de Chasseloup-Laubat would race each other five times in the following four months, taking turns in setting new records.[11]

Just a few years earlier, on 22 July 1894, another magazine, *Le Petit Journal*, had organized a Paris–Rouen race – often regarded as the first organized motor-car race[12] – with great success. What both *La France Automobile* and *Le Petit Journal* had discovered with these events was that racing made for great publicity – and thus more sales – and indeed the phenomenon was rapidly building an enthusiastic following. Races testing speed, skill and endurance were quickly established across Europe and North America. The first official Grand Prix took place in 1906 and was organized by the Automobile Club de France. It would set the tone for all motor sport to come, acting as a high-water mark of achievement for any car company and driver, although other races would gain similar mythical appeal, such as the *Mille Miglia*, a 1,000-mile race in Italy that ran from 1927 to 1957, and the *24 Heures du Mans*, a non-stop round-the-clock race in France, established in 1923 and still going today.

In the UK, fearing the lack of safety that racing at such new speeds would engender, the government passed a Motor Car Act of 1903, which put a ban on driving at over 20 mph on all public roads. Hugh Fortescue Locke King, an entrepreneur and race enthusiast, responded by building the world's first purpose-built racetrack for motor cars, Brooklands, which opened in 1907 [24].[13] The result was a site that became a national attraction for racing spectacle, hosting many early racing records, such as the first person to drive 100 miles in under one hour, which went to Percy E. Lambert on 15 February 1913.[14]

The site also became a contested space of gender politics. The Brooklands Automobile Racing Club banned women from competition in its early years.[15] However, this did not stop female drivers from using the track outside competitions and for other club races. Since its inception motor sport had included female participants, with drivers such as Camille du Gast in France and Dorothy Levitt in the UK competing side by side with their male counterparts at some of the earliest recorded races. Racing, and driving in general, formed

↑ **24** Colour offset lithograph poster advertising a motor race at Brooklands in October 1931. Brooklands was the first privately owned racetrack dedicated to motor sport in the UK, and host to a growing racing culture in the country

→ **25** A photographic portrait of Jill Scott Thomas, by Madame Yevonde, 1938. Thomas was a celebrated British race–car driver of the 1920s and '30s, active on the Brooklands circuit, where the participation of women in races had been historically contested

an important part of the growing women's rights movement – which would coalesce in the early years of the century around the suffrage movement – because driving was seen to be an empowering and liberating act. And so it was the image of women on the racetrack, conveyed at such events but also later through filmed documentation of the races, that turned accomplished drivers like Kay Petre and Jill Scott Thomas into powerful symbols of the movement [25].

Racing was an attractive subject for many turn-of-the-century photographers as well, who were drawn to it both by its visual richness and by the technical challenge of capturing such events. Jacques-Henri Lartigue was just a teenager when he took one of the most iconic racing photographs ever made: *Grand Prix of the Automobile Club of France, Course at Dieppe, 1912* [26]. In an attempt to capture the blurred field of vision experienced by a driver, Lartigue experimented by swinging his camera to follow the passing car as

he opened the shutter.[16] The frenzied framing and distorted figures, photographed during the 1913 Grand Prix of the Automobile Club de France, perfectly capture the experimental thrill of early racing culture.

Andrew Pitcairn-Knowles, on the other hand, focused his lens not so much on the cars as on the surrounding social spectacle of the race. From 1906 to 1908 he spent time in the Belgian coastal town of Ostend, where he documented the various people – and their attendant rituals and flamboyant displays – that accompanied motorsport events [27 & 28]. The series is an important snapshot of early racing culture, and of the rich elite who incorporated it into their conspicuous leisure routines, both as spectators and participants.

Races were not just about individual prestige, but also, crucially, about tests of industry. It cannot be overstated how much motor sport has driven the performance technology of cars throughout its history. Just as Jenatzy was trying to outdo not only his competitor, but also his competitor's brand – a Jeantaud – so too has almost every other car company sought to prove its worth on the racetrack. Henry Ford, who had a particular knack for storytelling and self-aggrandizement, briefly took up racing himself, after his first company went bankrupt. In 1901 he won a race in Grosse Pointe, Michigan, with a car of his own design, which ultimately helped him to secure funding to start a new company, and to go on to have the success that he did.[17] Ferrari has embedded the image of racing deep in its entire corporate identity, in part by mythologizing Enzo Ferrari's early racing career and the cars' multiple victories at Le Mans and in Formula One. Indeed, to this day some of the biggest brands still battle it out at Formula One and

↖ **26** *Grand Prix of the Automobile Club of France, Course at Dieppe, 1912*, photograph by Jacques–Henri Lartigue. Although titled '1912' by the artist, it was deduced later that the photograph must have been taken in 1913

→ **27 & 28** 'A Peugeot' and 'Getting Hooked', from a series of photographs taken by Andrew Pitcairn-Knowles in 1908 in Ostend, Belgium. The series offers a glimpse into the spectacle and fanfare of bourgeois motoring culture at the turn of the century

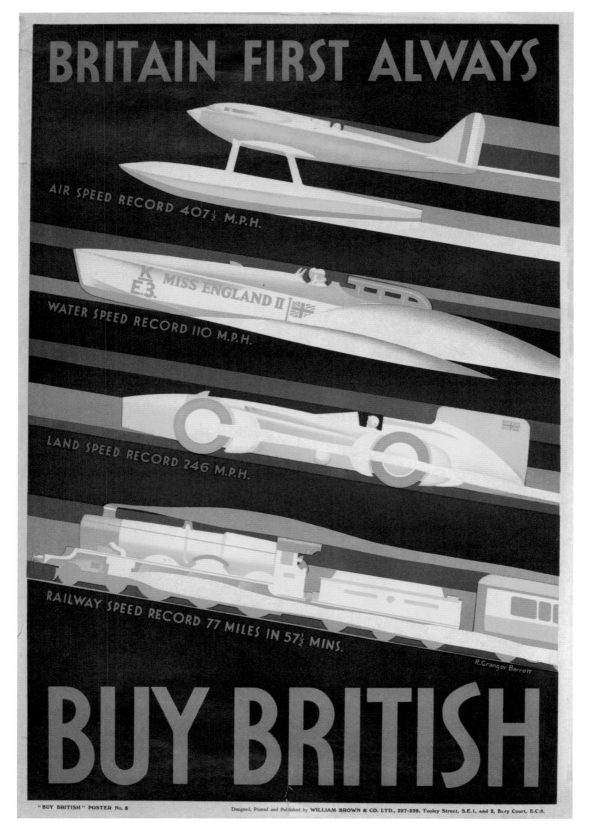

← 29 An Empire Marketing Board (EMB) poster, designed by R. Granger Barrett, c.1931, depicting a series of speed records held by British companies and individuals, including the Golden Arrow. Technical innovation was used as a promotional tool by the EMB to inspire people to buy more British products

→↘ 30 & 31 René Dreyfus racing and winning the Million Franc Race with his Delahaye Type 145 in 1937. The race was conceived as an incentive for French companies to build race cars that could better compete with the technologically advanced German cars of the era

Le Mans, the desire to be the fastest being spurred on by the priceless publicity that follows.

The might of industry has also been closely associated with the measure of a nation. And so, in the early history of racing, grand prizes and speed records were a matter of geo-political rivalry. The Empire Marketing Board used various British technological successes in its advertising campaign to 'Buy British First'. One poster included an image of the Golden Arrow, a car specifically designed to win back the land-speed record from the USA in the 1920s, taking the record at Daytona Beach in 1929 [29]. To do so, it was one of the first cars to experiment with the science of streamlining to minimize resistance from airflow, which has influenced car design ever since. Another example of such nationalistic bravado was the Million Franc Race in 1937 [30 & 31]. Organized by the French Popular Front, it offered a prize of one-million-francs to any French automobile maker who could produce a car fast enough to compete against the Nazi-backed Mercedes-Benz and Auto Union (later AUDI) drivers, who were dominating races at that time. The Delahaye Type 145 succeeded in the time trials held at the Autodrome de Montlhéry, taking the cash prize and ultimately validating the entire project when it defeated a legendary German Silver Arrow at the Pau Grand Prix in 1938.[18]

Speed Becomes Style

Meanwhile, by the 1920s and '30s, the public appetite for speed was not only helping to transform new driving technologies, but was also helping to fundamentally change the look of everyday objects. Most improvements to car performance were taking place under the bonnet, invisible to the naked eye, but in the first few decades of the century the application of aerodynamic testing began to have a drastic impact on the exterior shape of a car. From *La Jamais Contente*'s torpedo body to the Golden Arrow's (1928) low-slung, angular form, these early streamlining tests were mostly focused on one-off experimental cars designed purely to break records and, as such, for many years the aesthetic of streamlining remained relatively marginal. It was Hungarian-born engineer Paul Jaray, however, who sought throughout the 1920s and '30s to systematize the principles of streamlined car design so that they could be applied to modern production cars, thus bringing the look to a popular audience.

Working at Luftschiffbau Zeppelin in 1915, Jaray was able to use the company's wind tunnel to make his first major attempt at streamlined design, the airship LZ-120 *Bodensee* – whose bulbous front and tapered end would influence all subsequent zeppelin designs. In the

↓ **32** A French advertisement for the Tatra 77A, c.1937, referencing the elegant profile of its streamlined shape. The T77 was the first major opportunity for Paul Jaray – in collaboration with designer Hans Ledwinka – to implement his engineering study of streamlining into the design of a commercial production vehicle

1920s he used this knowledge to try and develop a comprehensive set of principles for streamlined car design, setting up a business to license such designs to automobile manufacturers.[19] While Jaray carried on discussions with many major car companies, and in some cases even developed test models, the Czech company Tatra was the only one finally to mass-manufacture a streamlined car in collaboration with him. After working with Tatra chief designer Hans Ledwinka, the T77 was revealed in 1934;[20] it was sleek, low and incorporated a fin running down its back [32]. Other streamlined designs would follow: the Chrysler Airflow debuted in the same year, with some dubious similarity to many of the ideas Jaray had been trying to sell to the company over the previous few years. Peugeot was a major adopter of streamlining methods, most markedly in its 402, which was launched in 1935, and its 202, a supermini that launched in 1938, with its characteristic sloping front grille. Perhaps most conspicuous, though, was the influence of the T77's streamlined successor, the T97, on Adolf Hitler and Ferdinand Porsche. Hitler specifically named Tatra as an example to follow, while directing Porsche to design him a people's car, and the resulting KdF-Wagen – the Beetle – bears many resembling traits.[21]

Meanwhile, the enthusiasm for streamlining, and the characteristic teardrop shape that could be found in many of Jaray's early aerodynamic tests, were already taking hold in the world of product design. American designer Norman Bel Geddes was a major proponent of the style, embracing it as what he thought was the wave of the future.[22] He used the teardrop shape in many automobile designs that were incorporated into his GM Futurama model for the 1939 World's Fair, in which he imagined the future of mobility in America. Raymond Loewy was also banking on streamlined style as some kind of natural evolutionary step. His 1930 'Chart of the Evolution of Design' depicts the morphing silhouettes of a car, a train, a telephone, a clock, a chair, a stemmed glass, a dress and a bathing suit. With each object he argued that over the course of time their essential shapes had improved and evolved, towards the smooth, flowing minimal forms characteristic of his era, giving streamlining a kind of pseudo-scientific Darwinian gravitas [33].

Of course, that a scientific study of aerodynamics – applied to zeppelins and automobiles for reasons of speed, efficiency and handling – would be translated to a world of static objects that were hardly in need of going anywhere fast is one of the great comedies of twentieth-century design. Soon, anything and everything that could be styled with sleek curves and tapered ends, to signify modernity and futurity, was given such a treatment. Pencil sharpeners, fans, irons and even meat slicers were formed as if they had been put through wind-tunnel testing, reducing what was a serious scientific study to a cheap marketing ploy [34, 35 & 36].

Streamlining eventually fell out of fashion, but the tendency to translate technical advancements in speed into other forms and mediums still remained. An equally misguided episode took place immediately after the Second World War, when Harley Earl, head of

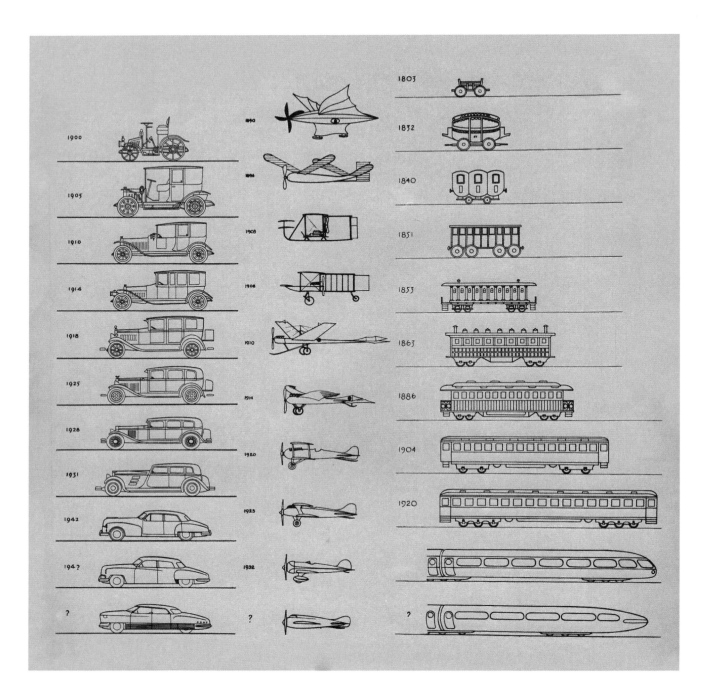

General Motors' Styling Department, became deeply inspired by the image of fighter jets and started translating their look into the design of his cars.[23] He began with a series of concept cars – Firebird I–IV – that were powered by jet engines and were tricked out with all the latest in aviation control technology. He soon found ways to imbue production cars with the look of cutting-edge flight, by including tail fins (ostensibly for stabilization at high speeds) and 'dagmar'

↑ ↗ **33** 'Evolution Charts', by famed American product designer Raymond Loewy, 1930, from a series illustrating how the design of several different types of objects had evolved over time to a more streamlined shape

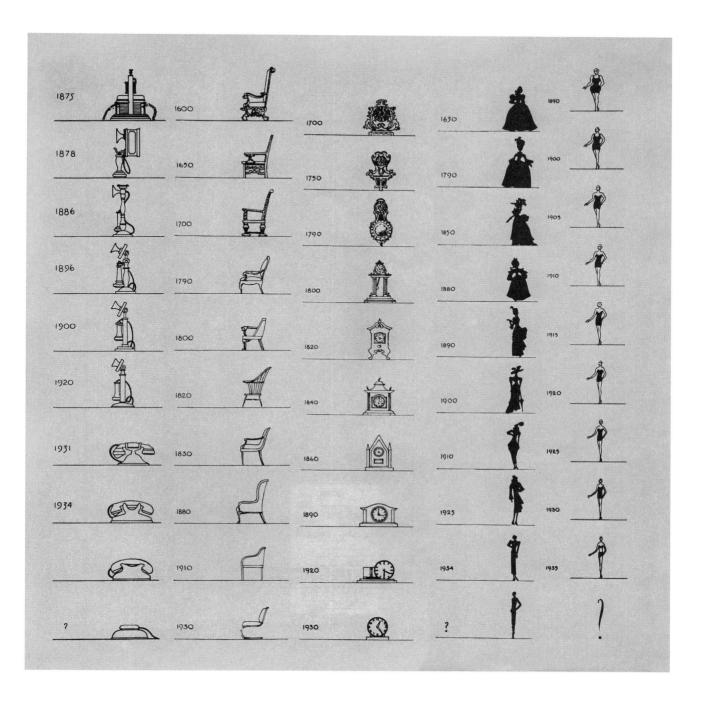

bumpers (to convey artillery shells or projectile missiles) on his Cadillacs, along with an excessive use of chrome. None of this had anything to do with the actual speed or functionality of the car, and as these accessories grew with each subsequent model, it became clear that styling had entered an aesthetic dead end, in need of a rethink.

Such follies speak, ultimately, of the restlessness of the public to go faster – and the response by car manufacturers to convey the

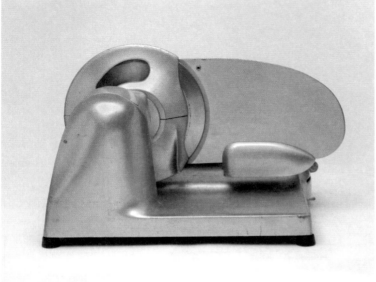

← **34** 'Airflow' table fan, by Robert Heller, 1937

↙ **35** 'Streamliner' meat slicer, by Egmont Arens and Theodore C. Brookhart, made after 1944

→ **36** Cloche hat, made by Miss Fox, 1928-9

By the 1930s and '40s, the science of streamlining had attained popular appeal, influencing the design of commercial products, including fans, hats, chairs and even meat slicers

image of accelerated speed directly in the car's design. But endless acceleration is of course impossible, and this demand for speed would soon encounter a crisis in the 1960s, which would change the car industry for ever.

→ **37** Advertisement for the Pontiac GTO Judge, 1969

Crash and Burn

By the 1960s the car industry in America was in need of new ideas. The stylistic excess of the tail-finned, chromed-up 1950s land cruisers had run its course. It was no longer enough to give people simply the illusion of speed by cross-dressing cars as jet fighters. Instead, a significant demographic – skewing young and male – desperately wanted the thrill of real speed emanating from under their car bonnets. The sleek and exotic sports cars they desired, however, were far beyond their price range. By 1964 the industry's answer to the problem could be found in two iconic cars that debuted that year: the Ford Mustang and the Pontiac GTO [37]. The idea uniting both was relatively simple: package all the power and performance of a classic sports car into something that was halfway affordable, essentially putting a big motor into a cheap mass-produced car. The GTO's name even alludes to that ambition by being compared to its sportier European cousins – *Gran Turismo Omologato,* a direct reference to the Ferrari 250 GTO.

These so-called 'muscle cars' quickly captivated the market, with each car brand racing to put out its own variation on the theme: the Oldsmobile Cutlass, the Plymouth Road Runner, the Chevrolet Camaro. Muscle cars became the quintessential Hollywood sidekick to the brooding male (anti-)hero particular to that era: Steve McQueen throwing his Mustang Fastback over the hills of San Francisco in *Bullitt* (1970), or Barry Newman popping speed pills as he raced his Dodge Challenger across America in *Vanishing Point* (1971). There was a pure hormonal recklessness to the idea of these cars, a nihilism mirrored in the late 1960s' general malaise brought on by the vicissitudes of the Vietnam War, protracted political protests and high-profile assassinations. Such recklessness sold. The Mustang brought record sales to Ford, selling 549,436 in 1966 in its peak year, and millions ever since.[24] It also prompted a horsepower arms race, in which each brand attempted to outdo its competitor with bigger, more powerful engines.

The problem with all of this was that fast cars translated fairly directly to human fatalities. And while the 'devil-may-care' attitude of moody muscle cars was good for movies, and for car sales, it was bad for litigation. High-profile crashes had punctuated the recent past – most notably James Dean in 1955, and Jayne Mansfield in 1967 – registering deep in the public consciousness the inherent danger of driving. Italy's famed *Mille Miglia* race was ultimately banned, after 30 years of operation, when two fatal crashes occurred in 1957. One involved the driver Joseph Göttgens outside Brescia, but the second, in the village of Guidizzolo, was far more horrific, killing

Born great.

Did you expect less? Shame! This is The Judge. And The Judge claims Pontiac's great GTO as its closest of kin.

Which explains The Judge's bump-proof Endura snoot. And the bulging hood scoops which can be opened or closed from the driver's seat. And the very unspongy springs and shocks. And the Morrokide-covered front buckets. And the no-nonsense instrument panel.

Now, if you want to think of The Judge as Billy's kid brother, OK. Just keep in mind that the family resemblance only goes so far.

You see, The Judge comes on with a 60″ air foil. A

custom grille. Big, black fiber glass belted tires. Special mag-type wheels. Blue-red-yellow striping. And name tags (like the wild one below) inside and out.

Keep in mind also that this baby performs like nobody's kid brother. Not with a standard 366-hp, 400-cube V-8 and Ram Air. Or a 370-horse Ram Air IV, if you so order. Either couples to a fully synchronized 3-speed with a Hurst shifter. Or order a close-ratio 4-speed. (Little old ladies might even order Turbo Hydra-matic.)

No sir. The kid brother hasn't been born yet that's greater, or tougher than The Judge.

THE JUDGE
A SPECIAL GTO BY PONTIAC

Four color pictures, specs, book jackets and decals are yours for 30¢ (50¢ outside U.S.A.). Write to: '69 Wide-Tracks, P.O. Box 888B, 196 Wide-Track Blvd., Pontiac, Michigan 48056.

not only the driver Alfonso de Portago and the navigator Edmund Nelson, but also nine spectators, five of whom were children.[25] Just two years earlier, an even greater tragedy had struck at Le Mans, when a Mercedes 300 SLR driven by Pierre Bouillin crashed and ignited, sending burning debris hurtling through the stands, killing 83 spectators and injuring 180.[26] By 1957 the increase in spectator deaths at such races was so dire that the Automobile Manufacturers Association, representing all the major car companies in America, agreed to withdraw all support from motor sport.[27]

While car-related deaths per capita had actually been steadily declining over the decades – due to better car and road technology – car companies were being held more accountable for the consequences of the speed they were selling. In the first half of the century, by contrast, manufacturers had largely been able to duck such finger-pointing, as the issue of car safety was framed around the responsibility and skill of the driver. Echoing the refrain used by gun advocates today that 'guns don't kill people, people do', early motoring advocates espoused the idea that cars were not the problem, but bad drivers were.[28] Awareness and safety campaigns were thus primarily directed at drivers, in order to promote better driving. In the UK, for example, the Ministry of Transport, as well as charities like the Royal Society for the Prevention of Accidents and the Pedestrians' Association, employed some of the country's most accomplished graphic designers and poster artists, such as Abram Games, Ashley Havinden and Tom Eckersley, to brandish posters with slogans like 'Obey the Speed Limit', 'Stop = Thinking + Braking' and 'Keep Death Off the Road' [38].

The book that changed everything in this regard was *Unsafe at Any Speed,* written by the young lawyer Ralph Nader in 1965. Nader took direct aim at the auto industry, castigating it for resisting the implementation of safety standards and, in some cases, engaging in dangerous cost-cutting exercises and improper safety testing. The book singled out the Chevrolet Corvair, which had a tendency to oversteer, causing a growing number of fatalities on the road. General Motors attempted to shift responsibility to the driver, selling optional 'stability enhancement' kits and requiring them to monitor their tyre-pressure differentials to correct the issue. By 1965, 106 liability lawsuits had been filed against General Motors because of the Corvair.[29] The book was vicious in its critique of the car industry and became a bestseller. As a result, just one year later, the US Congress passed the National Traffic and Motor Vehicle Safety Act, which for the first time established mandatory federal safety standards in the country. In launching a successful case against the auto industry, Nader had also helped to bolster a growing consumer movement, where grass-roots initiatives held producers to account for the products they manufactured.[30]

→ **38** A road safety poster issued by the Pedestrians' Association in the 1930s. As road deaths escalated due to increased automobile use, posters like these were published in the UK advocating for more responsible driving

The Shape of Us

The entire history of automobile design can, in part, be characterized by one enduring tension: between the individual desire to go fast and the collective desire to be safe. While speed had captivated much of the consumer's attention for the first half of the twentieth century, after Nader's success in America the tide shifted towards safety. This was reflected primarily in the design of cars, with a slew of innovations being introduced that fundamentally changed their look and performance, in the name of saving lives. In 1951 the first crumple zone was patented by Mercedes, allowing strategic parts of a car body to absorb the impact of a crash.[31] Three-point seat belts were invented by Volvo in 1959 [39], and from the 1960s onwards countries gradually enacted laws mandating that cars come equipped with them, and that people wear them.[32] Braking systems also improved vastly; by the late 1970s Bosch was perfecting anti-locking braking technology, which would subsequently be rolled out in almost every major car brand.[33] Airbags were first introduced in cars in the 1970s, and by the 1990s they were broadly legislated into manufacture. Today's cars are replete with safety features, many of which have gone digital – collision warning systems, lane-keep assist, electronic stability control, adaptive cruise control, 360-degree cameras, drowsiness alerts – aided by the advances in computer technology over recent decades. Driving has thus become drastically safer. In most countries, car-related deaths per capita are a fraction of what they used to be. In 1964, in the USA, there were 24 deaths per 100,000 people. By 2016 this figure had dropped to 11.6.[34] In 1966, in the UK, the total number of road casualties was 7,985. By 2016 this figure was 1,792.[35]

Road accidents are still, however, a leading cause of death across the world. And while they are tragic, having been with us for more than a century they have become normalized. Road deaths register to a certain degree as an unfortunate yet inevitable trade-off against the liberties and opportunities that mass car mobility provides. This idea of normalization is rendered into surreal and uncanny proportions by 'Graham', a project commissioned by the Transport Accident Commission in Australia and made by artist Patricia Piccinini in 2016. At first glance 'Graham' resembles a science experiment gone awry, a distorted figure with a vague human resemblance [40]. Instead, Piccinini attempts to chart a natural evolution of things, like Loewy's streamlining diagram, but this time not of the car, but of humans as we react and change in relation to the car. Essentially, 'Graham' projects how a human would need to evolve in order to survive a car crash: he has a flat face to absorb impact; his enlarged skull contains

↑ **39** A demonstration of Volvo's three-point seat belt. The company invented the seat belt in 1959, and subsequently introduced seat belts into their commercial vehicles in 1963

→ **40** 'Graham' is a recent project by Patricia Piccinini in collaboration with the Transport Accident Commission in Australia. It imagines what humans might look like, if they were to evolve to be able to survive car crashes

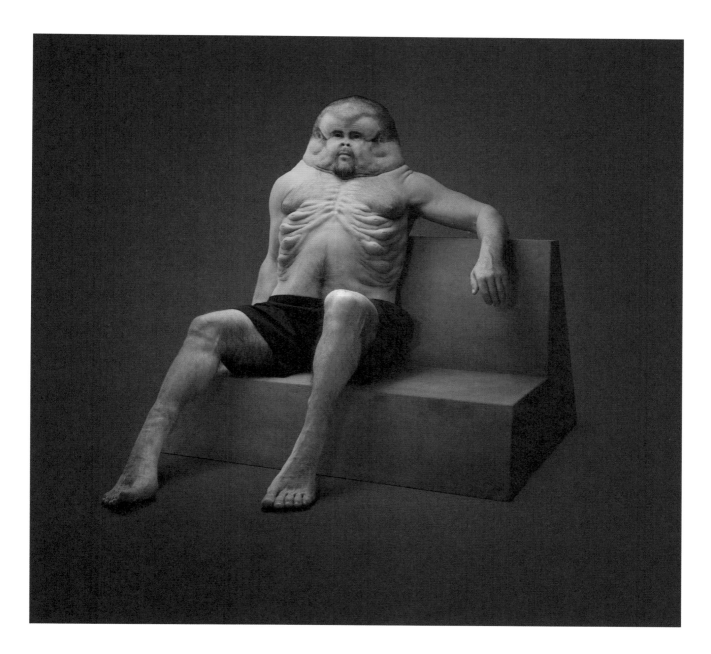

more fluid and ligaments to protect the brain; and the numerous nipples on his chest act as a type of airbag. 'Graham' embodies, in sensational over-the-top fashion, the classic Churchillian quip – 'we shape our buildings and afterwards our buildings shape us' – in a perfect synopsis of more than a century of designing for speed. We shaped speed, and now speed shapes us.

Oli Stratford

IN MY
FORD FIESTA

I don't think my 18-year-old self would have liked the journalist Ralph Nader. At that point in my life I hadn't read *Unsafe at Any Speed,*[*] Nader's 1965 exposé of the automobile industry, but I can only imagine how I might have reacted to its opening line: 'For over half a century the automobile has brought death, injury and the most inestimable sorrow and deprivation to millions of people.'[1] Well, sure, Mr Nader, but how about *you try living in rural Cheshire with no way of getting anywhere, and only the British Salt factory and field after field of dairy cows for company. What about* my *inestimable sorrow?*

Try to understand my tragedy. The nearest town of any real size to my house was at least half an hour's drive away. To get there, I could rely on my parents for a lift – although they both worked[†] – or else catch the only bus whose route passed anywhere close to my house. That bus took over an hour to reach its destination. It wound its way through blackberry-jewelled hedgerows and forest roads, stopping at what must have been every village within the county to pick up its crew of teenage rural goths and the accompanying retirees who peered at them in bewilderment.

Passing my driving test proved a godsend, and one whose divinity I was quick to cheapen by primarily using it to make trips to the petrol station to buy Twix – a form of profane mundane, which seemed a poor use for a mode of locomotion that is more regularly described with a sense of quasi-spiritual transcendence. In 1956, for instance, the Ford Motor Company put out a pseudo-study of road systems titled *Freedom of the American Road*. The project's promotional film saw the company's president Henry Ford II declare himself concerned with 'a special kind of freedom, a unique freedom' (the freedom of the American automobile owner 'to come and go as we please in this big country of ours'),[2] and this sense of unprecedented motion has long served as the founding mythology of the automobile. It's what Sal Paradise, Jack Kerouac's romanticized *roman-à-clef* nomad from *On the Road*, deemed the 'one and only noble function of the time';[3] and what Norman Bel Geddes – the American industrial designer whose 1940 book *Magic Motorways* envisaged a nation organized around automated highways – deemed the culmination of a human instinct to 'reach out farther and to communicate with other men more easily and quickly … a magnificently full, rich life for the people of our time'.[4]

Curiously, however, this kind of unfettered mobility has always seemed very far from my own experience. Perhaps mid-century America was simply too different a place from west Cheshire in the 2000s for any meaningful comparison to hold, but my perception of driving bore little resemblance to Kerouac's meandering existentialism, Bel Geddes's radical connectivity or Ford's pioneer-spirit expansionism. I wasn't roaming, communicating with others or discovering myself on the open road. Instead, the teenage me was mostly concerned with razzing up and down canal roads, listening to Kate Bush records on the CD player.

For me, and many of my peers, the promise of driving wasn't transformative in the fashion that Ford and his ilk had envisaged. Owning a car[‡] didn't radically expand my horizons or change the way in which I interacted with my surroundings. Rather, it led to a kind of densification of my existing world order. Once I began driving, I rarely ventured to new towns or took new roads. Instead I stuck to old routes that I knew intimately, and which I had been driven down by my parents throughout my childhood. Frequently the point of my car seemed less about travelling wherever I wanted, and more about travelling *however* I wanted. The ancient byways of the Stratford family had been opened up, and I – a souped-up Simba in a Ford Fiesta – was ready to express my mastery of them. I would evince control over the ancestral migration routes in the only way that seemed to make sense: by driving down them faster than my mum or dad did.

[*] Actually I still haven't read it in full. Don't like it; don't want to.

[†] And also had better things to do than driving me to Chester, so that I could sit pointlessly outside the Mozimo shoe shop with friends, until the owner came out and told us we were loitering and should reassess what we felt made for a memorable Saturday afternoon.

[‡] A baby-blue Ford Fiesta. I named him Trevor and he has since been sold. I understand that he is doing well in his new home.

In the early 1990s videogame communities began experimenting with the idea of the speedrun: a genre of play in which you complete a game in as short a time as possible. Games intended to last for 60+ hours are reduced into madcap half-hour dashes, all made possible by the players' mastery of the games' systems and maps. A classic of the genre is the rocket jump, a move by which players accelerate through an area by means of a process of self-detonation – lily-hopping ever onwards by means of the blast radius of rockets. As far as I can tell, the rocket jump was initially an unintended consequence of videogame physics engines, but this seems well within the spirit of the enterprise. Speedrunners are those players so familiar with a game that they can exploit and manipulate its mechanisms for their own ends; those for whom it's all so easy that they can just breeze through obstacles, bending and compressing time to suit their needs. This is as concise an explanation of my teenage driving style as I can provide.

This sense of self-inflation, of course, is the point. Driving offers speed as a commodity and, as with all forms of commodity, its allure is connected to the connotations that ownership of it suggests. US President William Howard Taft likened driving at speed to 'atmospheric champagne', which seems to capture perfectly the sense of giddy, heady thrill that is palpable within motion. The trick, however – at least outside J.G. Ballard's *Crash* (1973) – is that this stream of champagne must be safely directed and channelled, if it's to have its narcotic effect. The best drinkers are those who can hold their atmospheric champagne and not end up hiccoughing on the White House floor, just as the best drivers (or so the world of motor sport would have us believe) are those who can travel at such speed that the hazards of the route seem no obstacle to progression.

In her essay 'Contexts of Control', historian Miriam R. Levin has argued that contemporary usage of 'control', as related to 'mastery of the self', is linked to an earlier usage centred around industry – 'the apparatus by means of which a machine, as an airplane or automobile … was made to perform'.[5] This seems critical. If control operates as dominion over internal or external forces, it is a reasonably small leap to imagine that the greater those forces are, the greater the kudos attendant on their subjugation. It's the same principle behind why it seems more formidable, for instance, to walk a slathering pitbull terrier on a leash than a jaunty chihuahua. Certainly legislative perspectives adopt a framework of speed as something that must be tamed, if it is to have any utility. Police drivers, for example, are encouraged to travel quickly so as to meet the demands of their job, but equally to conceive of speed as something to be administered as a dosage within safely prescribed limits. 'It is emphasised that speed will never take precedence over safety', notes the publication produced following the 2009 National Police Driving Schools Conference. 'Drivers must demonstrate that they make reasoned and justifiable decisions to exceed a speed limit and that the speed they use is safe and proportionate to all the existing circumstances.'[6]

My early impressions of speed were not so sage. Instead they were shaped by my driving instructor, a motorcyclist whose creed of careful road stewardship was cut through with a seam of bravado tailor-made to worm its way into the skull of a teenage boy. *Sometimes people go past me on the straights*, he'd explain as we drove through the Cheshire countryside, stalling at every red light and accidentally setting the windscreen wipers furiously beating whenever we tried to indicate for a left turn. *And that's fine. They're breaking the speed limit and driving dangerously, but then I can kill them on the corners. I can go into those at 60 mph, and they can't believe it.* This seemed to me the epitome of mastery – an ability to make the difficult seem effortless – and so I came to harbour the belief that manoeuvres that seemed unsafe at any speed might, in fact, be broken and yolked. For those properly initiated into the mystery cult of the British School of Motoring, *every* speed was possible. Properly controlled, a Ford Fiesta might be a Tardis, a device for infinite motion whose only limitation was the aptitude and proclivities of its driver. The fact that my instructor also told me horror stories about a serious accident he had been in, losing his foot as a result, seemed to pass me by. Speed was for those in control, and control was what I aspired to.

This link between speed and control is, however, unstable. The rise of automated vehicles promises to remove direct control from the hands of the driver, or else reframe the idea of control from the Levinian notion of skilful utilization of a mechanical apparatus, into something more akin to easy consumption. Instead, it creates a situation in which vehicles, regardless of who sits at the controls, operate as a kind of service industry. 'Getting in a car will be just like getting in an elevator', promised the Tesla CEO Elon Musk at Dubai's World Government Summit in July 2017. 'You just tell it where you want to go and it takes you there with extreme levels of safety, and that will be normal.' Speed, in other words, will be reduced to a metric – one of many parameters locked into formation within the overall driving algorithm. One corollary of this change is that speed will seem neutered, forcibly uncoupling it from the machismo-soaked vision that it might ever have been a marker of expertise or dominion. In the world of the automated vehicle, speed will by nature be 'proportionate to all the existing circumstances', because this parameter will be designed into the system. To quote the design critic Stephen Bayley, 'the sordid, hot, dirty, dangerous romance of a car will be over'.[7] Or so runs the promise, anyway.

The problem, of course, is that any romance surrounding speed has always been based on childishness – a teenage phantasm cooked up as a justification for recklessness. In 2013 UK government figures showed that 3,064 people were killed or seriously injured in crashes where speed was a factor. Faced with the statistics, the mask of control begins to slip. Control over any speeding vehicle is, at best, *tenuous*. A driver's sense of dominion over their car is overinflated in its own terms, let alone when the vagaries and unpredictabilities of the road are taken into account. Movement at speed introduces variabilities that are difficult to accommodate, and which no number of protestations over control are likely to ameliorate. It was ever thus. Writing in 1928, the American economists S.H. Nerlove and W.G. Graham observed a 'definitely and consistently upward' trend in automobile activities, prompting them to ask, 'What can we expect from the automobile as an instrument of death in the near future?'[8] The answer, seemingly, was more of the same.

The mistake behind speed as control – the glitch in my Twix-rotted adolescent brain patterns – is the presumption that control betokens ease. As expertise increases, so too – we are led to believe – complexity resolves into simplicity. This premise, however, ought to be questioned. Writing in his *Futurist Manifesto* of 1909, the poet F.T. Marinetti eulogized speed as emblematic of progress and modernity. 'We drove on, crushing beneath our burning wheels, like shirt-collars under the iron, the watch dogs on the steps of the houses', he wrote. 'Death, tamed, went in front of me at each corner offering me his hand nicely, and sometimes lay on the ground with a noise of creaking jaws giving me velvet glances from the bottom of puddles.'[9] *Death tamed?* Today, I suspect, visions of progress in which death lurks in puddles feel anachronistic, as does the feverish optimism with which Marinetti felt speed would sweep all before it. Expertise may give us the tools we need to deal with complexity, but it does not flatten that complexity into nothingness. Those who drive well – maybe those who do *anything* well – account for, and accommodate, the obstacles in their path; they do not accelerate through them with abandon. 'I have one speed, I have one gear: go!', the actor Charlie Sheen once noted.[10] As adolescence has hardened into adulthood, so my conviction that Sheen is probably not right about a lot of things has only grown firmer.

MAKING MORE

Lizzie Bisley

THE MANUFACTURE OF PLENTY

→ **41** 'In Red Car to the
Kingdom of Communism',
a Russian revolutionary
poster, 1917–21

На красном автомобиле революции в царство
коммунизма.

A Russian revolutionary poster in the collections of the New York
Public Library shows a speeding figure driving towards us and out of
the picture, red flag flapping behind the car while an aeroplane flies
overhead [41]. Captioned 'In Red Car to the Kingdom of Communism',
the poster's car races out of the factory and ahead into a communist
future – cranes and conveyors soar above, while smoking chimney
stacks and giant factory buildings work steadily in the background.[1]

Industrial production defined, in many ways, the course and shape of the twentieth century. This poster highlights the extent to which the manufacture of that most modern of technologies – the automobile – sat at the centre of this. Here the car is presented as a politically and economically transformative object, its speed and power indelibly linked to the factory that produced it.

Although it came from a very different end of the political and ideological spectrum, this Russian image of a new, modern world fuelled by industrial manufacture is deeply embedded in the capitalist revolution of American 'Fordist' mass production. From the 1910s, in reaction to a flood of affordable Model T cars pouring off the assembly line, manufacturers around the world raced to adopt the Ford Motor Company's production techniques. Outside the factory, the dream of 'Fordism' took on a life of its own – its principles adopted as the means of creating a new kind of society, or a new way of living. Mass production, with its roots firmly entwined around the image of the cheap automobile, came to represent the promise of a brighter future for artists, writers, designers, politicians, economists and industrialists all round the world. Shaping their work in real and significant ways, it seeped deeply into both the consciousness and the physical shape of the twentieth century.

'A lever to move the world'[2]

The idea of the Model T as a world-changing car is now so familiar that it is hard to fully appreciate the gigantic shift that its production represented. Ford aimed to make a car for the 'masses' – simple, robust, easy to run and cheap. The first Model Ts were manufactured in 1908, and Ford continued to produce this single model (with some variations) until 1927, at which point it was replaced by the new Model A. The Model T experienced a meteoric rise on the US market, with the Ford Motor Company quickly surpassing the production of all its competitors: the car had taken over a giant 45.6 per cent of the US market within its first six years of manufacture.[3] By the time Ford stopped making the Model T in 1927, 15 million had been produced, which made it the most successful car in the world until as late as 1972, when it was finally outsold by the VW Beetle.[4]

As production sped up, the price of the Model T fell steadily. Starting in 1908 at a very affordable US$850, it had dropped to an astonishing $298 by 1923.[5] This represented a gigantic shift in the market: in 1907 the average price of American-manufactured gasoline cars was $2,834, while for European cars the figure was a startling $6,730.[6]

With this cheap machine on the scene, car ownership rocketed in the USA and the automobile became, for the first time, a commonplace and widely accessible means of transport. American car registration went from less than half a million in 1910 to 26.7 million in 1930. The Model T, with its sturdy, simple design, became the standard, all-purpose vehicle of tinkerers across the continent – adapted for

everything from a snowplough to a caravan [42]. Other companies race to match Ford in price and production, and approximately half of all American households owned a car by the mid-1930s.[7]

The Model T was also, almost immediately, a truly global car. By 1915 it was being manufactured in 28 factories around the USA, with a further 14 factories and distribution points in Europe, Canada, Latin America and Australia.[8] Articles published in a tiny New Zealand newspaper, the *Bush Advocate,* in May 1910 show how far and fast it had travelled. Commending the 'efficiency and speed-climbing capabilities' of these 'reliable cars', the paper noted that they were available from a local dealer.[9]

Lots of factors contributed to the speed and scale of the Model T's production, including the simplicity of the car's design; the use of standard parts to manufacture at scale; and the development of pressed steel techniques to produce parts more quickly. Most significant, however, was the Ford Motor Company's use of a sophisticated moving assembly line, a system designed to roll cars through the factory in a constant flow of manufacture.

The assembly line, which was first put into operation at Ford in 1913, essentially worked by breaking down the building of each car part into its smallest actions. Rather than workers taking a component through the various stages of its construction, a single person would perform each individual action, standing still and repeating this motion as an endless stream of parts moved past them along a conveyor belt [43].

The use of a moving assembly line hugely sped up Ford's rates of production. As each person in the system performed only one action, no time was lost moving around the workshop floor. Each piece of work was completely standard and regulated, often making use of a tool or machine developed for that specific operation. This eliminated almost all room for error, while the constant flow of parts kept people moving in a very fast, continuous rhythm. The standardization of the system, and the extreme deskilling of factory-floor workers, also meant that the Model T's manufacture could be relatively easily replicated across factories around the world.[10]

One of the most fascinating things about the assembly line is the intensity of its design. In order for the production of a complex mechanism like a car to be recorded, and then exactly repeated, each part's construction needs to be mapped down to the last turn of a screw (the Model T's construction involved 7,882 separate jobs).[11] This mapping is then laid out across both the physical space of the factory and the movements and bodies of the workers.

The density of design around the Model T's manufacture was expressed in part in the factory building, which came to operate as a giant cog in the production machine. The architect Albert Kahn

↖ 43 Assembly of the magneto flywheel at Ford's Highland Park factory, April 1913. After adding parts, each worker would push the flywheel along to the next person in the line

↗ 44 Craneway used to move materials and parts between levels of the Ford Highland Park factory, c.1914

designed Ford's Highland Park factory, which opened in 1910, in consultation with the mechanics who worked on the car. Conspicuously modern in its openness, light and ventilation, the factory also traced the assembly of the Model T – the parts of the car were gradually constructed and assembled as it moved from the top to the bottom of the building [44]. By the time Ford opened its second major Detroit factory, River Rouge, in 1927, operations had become even more refined. Here Kahn created a huge industrial city, with different parts of the overall manufacture streamlined into different buildings on a single site [45].

In its intensity of purpose, and its magnitude of production, the Ford assembly line sprang in part from the culture of experimentation that was fostered among the group of mechanics working at Ford in its early years.[12] On a larger scale, Ford's processes were also deeply embedded in the manufacturing culture of the nineteenth century, which had been strongly shaped by increasing industrial mechanization, a move towards interchangeability and scale production. This longer history of industrial process was something that Henry Ford himself acknowledged, claiming that the inspiration for the line came from the Chicago meat-packing industry, with the

→ **45** Flow chart showing the stages of production at the Ford River Rouge plant in 1941, from raw materials to assembled car

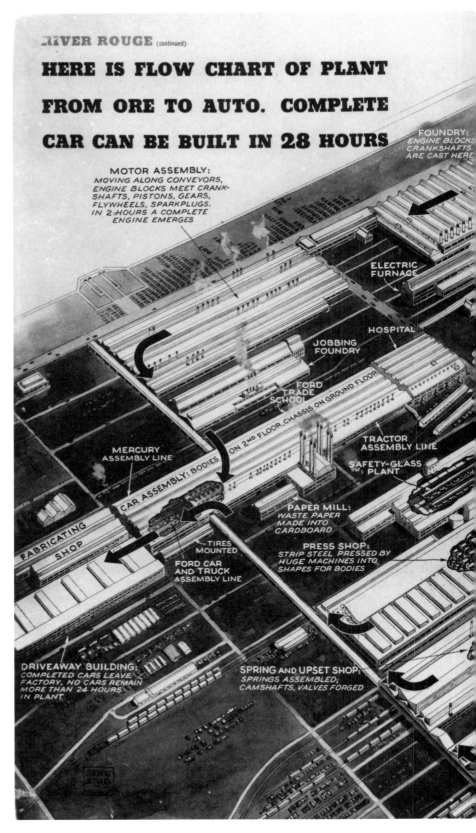

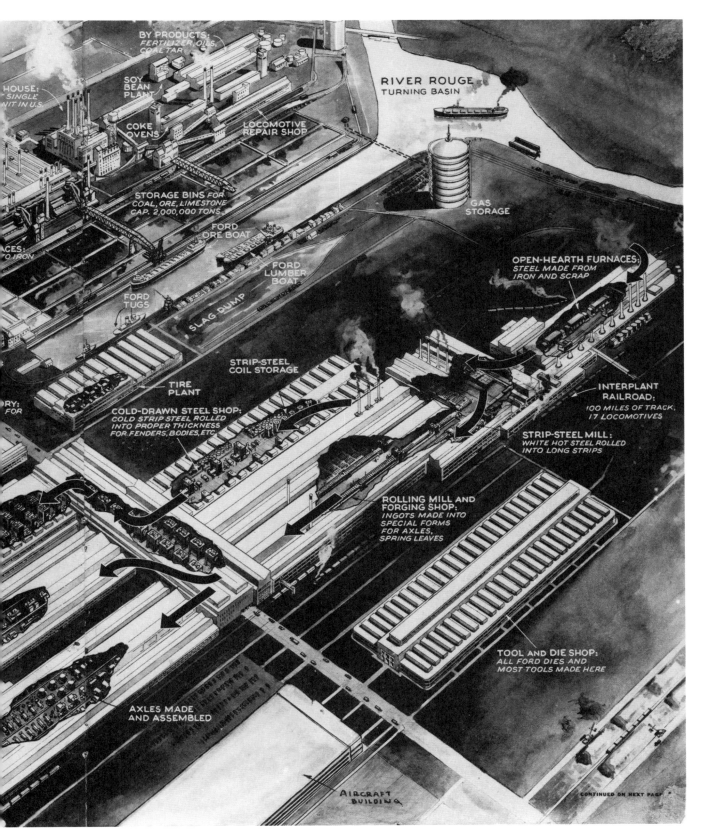

BY PRODUCTS:
*FERTILIZER, OILS,
COAL TAR*

SOY
BEAN
PLANT

HOUSE:
*SINGLE
UNIT IN U.S.*

COKE
OVENS

LOCOMOTIVE
REPAIR SHOP

RIVER ROUGE
TURNING BASIN

GAS
STORAGE

ACES:
O IRON

STORAGE BINS *FOR
COAL, ORE, LIMESTONE
CAP. 2,000,000 TONS*

FORD
ORE BOAT

FORD
LUMBER
BOAT

FORD
TUGS

SLAG DUMP

OPEN-HEARTH FURNACES:
*STEEL MADE FROM
IRON AND SCRAP*

STRIP-STEEL
COIL STORAGE

TIRE
PLANT

RY:
FOR

COLD-DRAWN STEEL SHOP:
*COLD STRIP STEEL ROLLED
INTO PROPER THICKNESS
FOR FENDERS, BODIES, ETC.*

INTERPLANT
RAILROAD:
*100 MILES OF TRACK,
17 LOCOMOTIVES*

STRIP-STEEL MILL:
*WHITE HOT STEEL ROLLED
INTO LONG STRIPS*

ROLLING MILL AND
FORGING SHOP:
*INGOTS MADE INTO
SPECIAL FORMS
FOR AXLES,
SPRING LEAVES*

TOOL AND DIE SHOP:
*ALL FORD DIES AND
MOST TOOLS MADE HERE*

AXLES MADE
AND ASSEMBLED

AIRCRAFT
BUILDING

CONTINUED ON NEXT PAGE

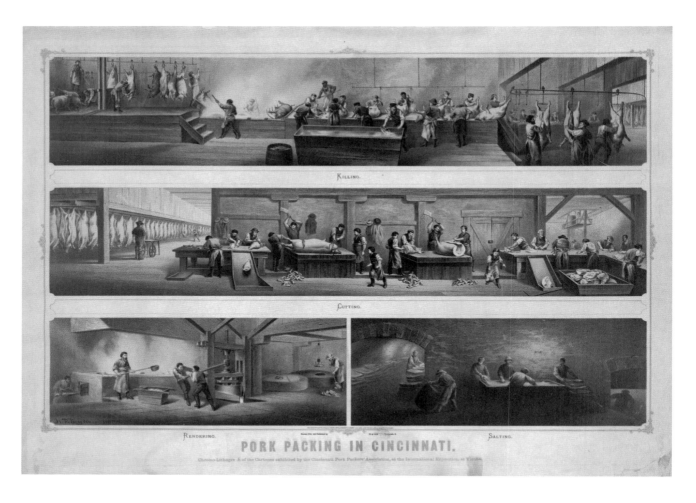

KILLING.

CUTTING.

RENDERING.   SALTING.

PORK PACKING IN CINCINNATI.

Chromo-Lithogr. & of the Cartoons exhibited by the Cincinnati Pork Packers' Association, at the International Exposition, at Vienna.

moving conveyor a direct adaptation of the overhead trolley used in industrial butchery to disassemble a carcass [46].[13]

More broadly, however, Ford's production techniques must also be seen in the context of the late nineteenth-century work of Frederick Winslow Taylor, an American engineer whose study of 'scientific management' sought to standardize the motions, processes and tools used by factory workers in order to make production more efficient and rational.[14] Taylor's ideas were developed by a raft of other engineers and writers, crystallized both inside and outside the factory in the work of figures like Lillian and Frank Gilbreth, who conducted influential studies on 'time and motion' that sought to eliminate wasted actions from domestic, industrial and work settings. Ford's system both sprang from, and came to personify, this much wider cultural climate of rational planning and efficiency.

What is remarkable about the Ford Motor Company is the extent to which it pushed these existing ideas to their furthest possible degree. Ford aimed to make the production of cars the central point in an entirely regulated and controlled universe. This would extend from the design and manufacture of car parts to the sourcing of

↑ **46** Chromolithograph describing the use of a moving conveyor belt by Cincinnati butchers in the 1870s, published by Ehrgott and Krebs, c.1873

→ **47** Fordlandia Estate Map, 1936. This vast rubber plantation covered 10,000 square kilometres of land in the Brazilian Amazon

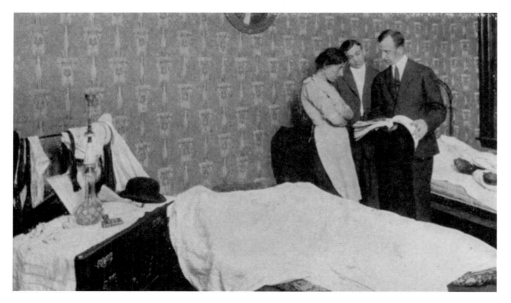

materials, the transportation of the finished car, a system of car dealers and a network of mechanics. He wanted to make every part of the car himself, a scheme taken to its most extreme with the establishment of a giant rubber plantation, 'Fordlandia', in Brazil in the late 1920s [47]. Although the plantation was a spectacular failure, Ford intended it to produce tyres without fear of shortages in supply or price rises.[15]

Ford sought to control the movement of Ford workers not only in the factory, but also in their homes and in their heads – most directly through the company's infamous 'Sociological Department', founded in 1914, which aimed to regulate and record the social habits and living conditions of factory workers and their families [48]. Lastly, and perhaps most significantly, he saw his Model T as part of a complete, circular economy: Ford aimed to turn his producers into consumers, manufacturing cars that workers could buy, and paying them enough so that they could stimulate demand in an ever-larger market of mass-produced commodities.

48 The Sociological Department gave workers prescriptive advice on domestic conditions, school attendance, household finances and hygiene. These photographs, taken in 1914 and 1915, show a Department employee inspecting conditions in the home of a Ford worker (top) and living conditions deemed 'undesirable' by the Company (bottom)

↘ **49** One thousand Ford Model T chassis, the production of a single shift, outside the Highland Park factory, August 1913

'The standardised cultural product is the hit tune'[16]

A photograph taken in 1913, the year that the moving assembly line began its full operation, shows 1,000 Model T chassis outside Highland Park – the output of a single shift's work [49]. Astonishing in its endless ranks of identical parts, the photo celebrates the vast scale, speed and standardization of Ford's manufacture. Widely distributed in books, as postcards and in the press, the image was a small part in a much wider campaign by the company to publicize and mythologize the bounty of the assembly line.

The Ford Motor Company spread news of its manufacture in a vast number of ways, including through detailed technical accounts aimed at industry, and in popular books such as the 1915 volume *Ford Factory Facts*. Thousands flocked to Highland Park for factory tours, which were produced alongside motion pictures (the company had a dedicated film crew on its staff), photographs, advertising, press and spectacular displays at World's Fairs: in 1915, at the San Francisco Panama-Pacific International Exposition, a working assembly line was erected to turn out a Model T car every 10 minutes for three hours of each day.

The Model T's remarkable success, when combined with this spectacle of production, worked to establish the Ford Motor Company as

the embodiment of twentieth-century methods of mass production. As a correspondent for the *New York Times* wrote in 1931: 'Fordism became a fetish for industrialists everywhere … The promise of mass production, the resultant economy in costs, the reduction in overhead, the ability to produce a good article at an unbelievably low price, became the basis of a new industrial cult.'[17]

By the early 1920s the word 'Fordism' had come into general use in the USA, Britain, Europe and Australia.[18] Ford's techniques were widely copied and adapted across the automobile and other industries. The shoe manufacturer Tomáš Baťa, for example – dubbed the 'Czech Ford'[19] – studied Ford's methods closely when establishing his own version of assembly-line production in factories around the world.

At the Fiat factory in Turin, Fordist mass production was firmly and directly embraced, with Giacomo Matté Trucco's Lingotto factory (started 1916, completed 1923) designed as a mirror image of the Highland Park process: raw materials were brought into the lowest level of the five-storey building, and the car was constructed and assembled as it moved up to the top. In a particularly thrilling twist on Ford's spectacle of assembly, finished cars drove – like magic – out of the factory and straight on to a test track on the Lingotto roof [50].

Alongside the eager adoption of Ford's processes by industry, the company's mass-production system also had a wider ideological influence. Throughout the 1920s and '30s the principle of standardized mass production – deeply embedded in the figure of Ford – came to be identified as one of the chief building blocks in the creation of a new world. One direct example of this can be found in a 1924 letter to Henry Ford from the former President of the Republic of China, Dr Sun Yat-sen. Stating that he was 'of the view that China may be the cause of the next World War if she remains economically undeveloped', Sun saw the establishment of Ford factories in China as a way of avoiding this fate.[20]

Expressing a similar belief in the power of standardized production to enact social and political reform, Ford's methods were also championed in Soviet Russia as part of post-revolutionary plans for a socialist, industrialized society. Under Stalin's first Five-Year Plan, an agreement was reached with the Ford Motor Company in 1928 to build a factory (designed by Albert Kahn) for the construction of Fordson tractors.[21] In 1929 the Soviet Union signed another agreement to build an assembly plant for the Ford Model A. Delegations

of engineers and factory workers travelled to Stalingrad to help with the start of operations, training the Soviets in the processes of assembly-line production. The Nizhni Novgorod plant, intended as a Soviet Detroit, began full production of the Model A in 1932, using equipment shipped from River Rouge [51].[22]

Alongside this direct economic investment, Henry Ford himself was an extremely prominent figure in Russia in the early 1920s. Not only was his 1922 autobiography published in eight Soviet-translated editions, but he also became a remarkably popular symbol for revolutionary change: there are accounts of workers in parades holding banners bearing Ford's name, while contemporary journalists claimed that Ford was better known in rural Russia than most communist figures, citing examples of peasants naming their children after him.[23]

Within urban planning circles, the words '*Fordizatsiia*' and '*Fordizm*' were adopted by Russian discourse, used interchangeably with 'American efficiency' to describe a system of material wealth that could be adapted to socialist aims.[24] As Sonia Melnikova-Raich has argued, despite their origin in capitalist enterprise, the Soviets considered both Taylorism and Fordism to be 'ideologically neutral' techniques that could serve the cause of communism as well as they had served capitalism.[25] Mass production, embodied in the figure and factories of Henry Ford, was seen to offer the path to a modern society of socialist plenty.

A similar idea was expressed by the American technology writer Lewis Mumford when he wrote about mechanized production in 1934:

> Whatever the politics of a country may be, the machine is a communist … the work represents a collaboration of innumerable workers … And the product itself necessarily bears the same impersonal imprint: it either functions or it does not function on quite impersonal lines …
>
> In money-ridden societies … every attempt is made to disguise the fact that the machine has achieved potentially a new collective economy, in which the possession of goods is a meaningless distinction, since the machine can produce all our essential goods in unparalleled qualities, falling on the just and the unjust, the foolish and the wise, like the rain itself.[26]

Mumford argued that the only way to truly express the collective potential of the machine age was through a machine aesthetic, in which the standard, mass-produced object was embraced over the stylized commodity.[27] It was a belief already made concrete in the work of European modernists, many of whom sought to embrace the power of standardized production as a way of making human relationships with objects more rational, progressive and universal.

These principles shaped the work of many influential twentieth-century designers: from Hannes Meyer, the second director of the Bauhaus, who argued that standardization of all things, from clothing to culture ('the standardised cultural product is the hit tune'),

↗ **52** Woman sitting in Marcel Breuer's tubular steel club chair, wearing a mask by Oskar Schlemmer, photograph by Erich Consemüller, 1926

would have a democratic, liberating effect, and who proposed social reform through cooperative mass production; to the Hungarian designer Marcel Breuer, who developed a series of standard chair forms in the late 1920s [52].[28] Although Breuer's chairs were not in fact mass-produced until much later, they were heavily steeped in a utopian vision of mass manufacture. Reducing the chair to its most basic, functional components, and using conspicuously industrial materials such as chromed tubular steel, Breuer described his designs in the language of factory production: 'All of the various types are constructed of the same standardized, elementary parts that can be disassembled and interchanged at any time.'[29]

The factory also became a model for the construction of the house, and for the work that took place inside it. Margarete Schütte-Lihotzky's so-called 'Frankfurt' kitchen, which was installed in thousands of German social houses, intended to create a more rational organization of domestic work – giving women the time, energy and space for other things [53]. The design was based on a close analysis of the movements of a housewife. In a manner similar to that of Ford's assembly-line mechanics, or a time-and-motion engineer, Schütte-Lihotzky mapped these out on to the footprint

of a room, designing a fitted kitchen whose appliances, sinks and work surfaces were configured in a way that would allow a meal to be cooked using the smallest possible number of actions.

This rationalization of domestic work was matched outside the house by an architectural obsession with the idea that systems of car manufacture could be replicated for the mass production of housing. Architects returned to this idea consistently, and with passion, from the 1920s to the 1960s, believing that if the power of the automobile assembly line could be harnessed in house production, this would mean the ability to solve easily and cheaply the housing crisis (or to sell a lot of houses, depending on which side of the ideological divide they fell on).[30] This idea was expressed at its most literal in the work of the German architect Ernst Neufert, who proposed in 1943, as a means of replacing housing bombed during the war, the construction of a 'Hausbaumaschine': a factory on rails, which would extract a long line of housing blocks as it moved forward through the damaged city [54].

↑ 53 Margarete Schütte-Lihotzky's 'Frankfurt' kitchen was designed to make domestic labour more efficient, rational and less time-consuming, 1927

→ 54 Cross-section of Ernst Neufert's *Hausbaumaschine*. The scaffolding frame was designed to move forward on rails, leaving a trail of fully assembled housing blocks behind it. Published in Ernst Neufert, *Bauordnungslehre*, 1943

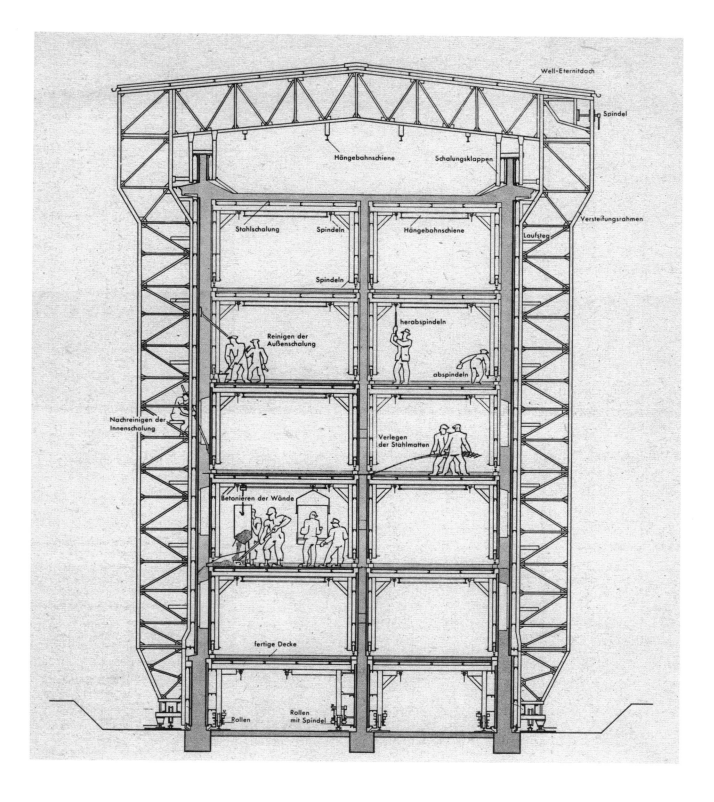

'The only thing they have on their mind is to keep that line running'[31]

Neufert's housing machine has an obvious and dark underside. Constantly moving and engulfing new space, it puts the never-ending production of the assembly line into the context of Nazism and the most violent systems of the twentieth century. The ideal of the universal standard has a counterpart in eugenics: a removal of the individual and a move towards an all-controlling machine. This flipside, in which mass production becomes a dystopian engine of mechanization and control, was expressed vividly throughout the twentieth century. The focus of much socialist and worker debate in the 1920s, it was an image that became increasingly popular and widespread from the 1930s. It can be seen in the 1931 French film *À nous la liberté*, in which the production line is literally situated in a prison [55]; in the unrelenting pace and control of the factory in Charlie Chaplin's *Modern Times* (1936); and in Aldous Huxley's 1932 novel *Brave New World*, which depicts a horrifying fascist universe, planned and controlled down to each person's last action and presided over by a god called Henry Ford.

These fictional worlds had a very stark reality in the grim conditions of working on the assembly line. In a letter to Henry Ford in 1914, the wife of a Highland Park worker wrote:

↓ **55** Still from the film *À nous la liberté* (1931), directed by René Clair, showing an assembly line used to manufacture phonographs

33,400 MEN ARE MADE IDLE; 'SIT-DOWN' STRIKES CLOSE 7 GENERAL MOTOR PLANTS

↑ **56** Workers at a General Motors factory that was closed by the 1937 sit-down strike in Flint, Michigan

The chain system you have is a *slave driver*! ... My husband has come home and thrown himself down and won't eat his supper – so done out! ... Couldn't there be a man ready to step in and relieve a man when nature calls ... That $5 day is a blessing – a far bigger one than you know but *ok* they earn it ... *Please investigate* ...[32]

From the assembly line's earliest days, concerns were raised by workers, unionists and socialist writers about the exhaustion, boredom and relentlessness of such work. The difficulty of being on the line was indicated by the massive increase in staff turnover at Ford – after it was introduced in 1913, annual turnover of factory workers went up 380 per cent, and it was this unsustainable rate that led to the introduction in 1914 of 'profit-sharing' schemes and the (short-lived, but incredibly highly paid) $5 day.[33]

Through the 1930s, in the wake of the Depression and on the back of unemployment, falling wages and poor working conditions, there were increasing attempts to unionize the US automotive industry. Sit-down strikes at General Motors plants in Flint in 1936–7 eventually led to the company signing a contract with the Union of Automotive Workers (UAW) [56]. This was the first large-scale victory for car unions in the USA and spurred similar strikes in plants around the

↑ **57** *Detroit Industry*, by Diego Rivera, 1932–3, a mural on the north wall at the Detroit Institute of Arts, showing the production of Ford's V8 engine

country – union membership went up from 30,000 to 500,000.[34] The Ford Motor Company led a particularly dirty and violent resistance to unionization, most infamously at the 1937 Battle of the Overpass, during which UAW members handing out pamphlets entitled 'Unionism not Fordism' were severely beaten by company heavies.

At the same time that strikes were raging through Detroit, and in the wake of the bloody Battle of the Overpass, the Mexican artist Diego Rivera was commissioned by Edsel Ford (Henry's son) to paint a series of murals for the Detroit Institute of Arts. The murals, which show the production of the new V8 engine at River Rouge, convey something of the hungry, all-encompassing movement of the factory. In the painting for the north wall [57], workers making engine blocks move their bodies alongside the moving parts, each of which is a

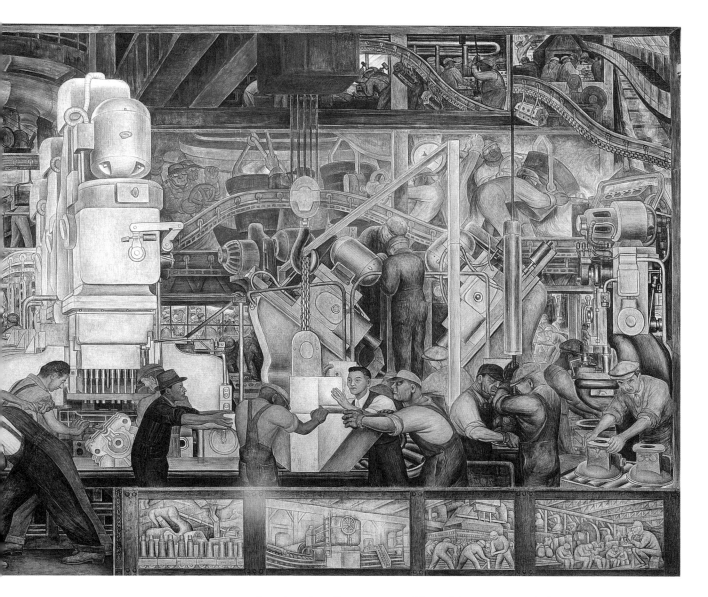

tiny piece in a fierce storm of foundries, conveyor belts and men.

This unrelenting, inhuman speed became a focus in labour-movement protests against factory conditions: one of the points raised in the UAW's 'Unionism not Fordism' leaflet was a wish to 'Stop Speed-up' of the line.[35] Although Ford's initial goal of manufacturing a single, standard car had been – by 1927 – displaced by General Motors' successful use of stylized market differentiation (see pp. 95–115), the assembly line's overwhelming aim to produce more things, more quickly, continued to dominate both the automobile and other industries. From as early as the 1920s companies looked to do this by increasing automation and mechanization on the line.[36]

Mechanization was introduced to car manufacture largely through the 'transfer machine' – a multi-headed tool that automatically

FIGHT AUTOMATION FALLOUT

WITH A SHORTER WORK WEEK

← **58** Union of Automotive Workers' poster protesting against factory automation, 1950s

→ **59** Unimates working on the assembly line at General Motors' Lordstown factory, photograph by Michael Mauney, *Life* magazine, May 1972

performed a number of operations on a part, before moving it along to the next machine. Transfer machines were hugely productive and came to be seen as key to maintaining high (and highly politicized) levels of US manufacture during the Cold War.[37] They were also widely criticized by unions for speeding up production, reducing pay and increasing unemployment [58].

By the 1960s the transfer machine was, however, already in the process of being replaced by a new kind of machine, one that offered far more flexibility as it could be programmed to perform a range of operations. The first industrial robotic arm was developed by the inventor George Devol, who applied for a patent in 1954. The prototype went into use in a General Motors die-casting plant in 1959.[38] Dubbed the 'Unimate' by Devol and his business partner, Joseph Engelberger, 450 of these arms were put into the General Motors factory in 1961, while the technology also quickly spread outside the car industry: Nokia was licensed to manufacture the robots in Scandinavia; and the motorcycle company Kawasaki manufactured them in Asia.

In 1969 General Motors rebuilt its three-year-old plant at Lordstown, Ohio, introducing Unimates to do 95 per cent of all welding on the line [59]. The Lordstown plant, lauded in the press as the factory of the future, was devoted to manufacturing a new, small car – the Chevrolet Vega. By using Unimates, simplifying design and introducing new work rules, Lordstown increased its production from 60 cars per hour in 1966 to 100 by 1969.[39] This made it one of the most productive plants in the world, and the company was so aware of the crushing speed at which it was operating that it purposefully hired younger workers, in the hope they would be able to match the pace of the line.[40]

In spite of the rhetoric about a revolution in manufacture, things quickly started to go wrong at Lordstown. By 1972, UAW Local 1112 had shut down production at the plant in protest at the unreasonable speed. As Gary Bryner, local union president, recalled in 1974: 'If the guys didn't stand up and fight, they'd become robots too ... Thirty-five, thirty-six seconds to do your job ... Go to the next job with never a letup, never a second to stand and think.'[41] Although the Lordstown shutdown was effective in the short term, leading to a return to 1966 working conditions, the takeover represented by the 'praying mantis' Unimate was ultimately here to stay.[42]

At the same time as General Motors was changing up its production with the introduction of robotic arms, the global automotive industry was becoming increasingly dominated by small, affordable Japanese cars (the Chevy Vega had in fact been designed to compete with this manufacture). Japanese cars were not only popular for their size and fuel-efficient design – particularly desirable after the 1973 oil crisis – but were also manufactured in increasingly efficient factories, which enabled them to be sold at very reasonable prices. Starting in the 1950s, Japanese car companies began to outstrip the big US manufacturers in the relative productivity of their plants. By the late 1970s and early '80s, the number of cars produced by each Japanese worker in a year was roughly twice that manufactured in the USA.[43] The fundamental reason for this was the adoption of a system of manufacture that has since been dubbed 'just-in-time', which was developed at Toyota from as early as the 1930s.[44]

While other Japanese manufacturers, such as Nissan and Mitsubishi, initially produced cars under licence with major American and British companies, Toyota could not afford to buy American production systems when they began operation in the 1930s. In response to this they developed a system that was based on US mass production, but remained much more flexible to the vagaries of demand. A large variety of parts was sourced from other suppliers, and the supply of materials and parts operated on a system known as *kanban*: essentially meaning that you would only produce parts as needed, signalling the rise and fall of demand through a simple exchange of paper tags. Under just-in-time manufacture, all workers were encouraged to stop the line if they became aware of problems, meaning that production became much more decentralized and responsive.

Although Toyota's system was born from a study of American mass-production techniques, in many ways it turned those principles on their head – rather than having a river of constantly flowing goods, Toyota (and the other Japanese manufacturers that adopted the system) positioned itself as a stopping and starting tap, quick to flow when turned on, but also quick to turn off when there was danger of a flood.

In the same way that Fordism did in the 1920s, just-in-time changed the landscape of manufacture once again, heralding a giant shift towards global supply chains, manufacture on demand and flexible production. Where once Ford had imagined creating an entirely controlled universe, working out from a single point with every action planned, the rise of just-in-time reflects the much less centralized, more uncertain conditions of the late twentieth and early twenty-first centuries. When combined with the long and ever-more dominant afterlife of Devol's 'Unimate', and the gradual spread of robotic production across different kinds of manufacture, we find that the very particular conditions of early twenty-first century industrial production have their roots firmly lodged in the car manufacturers' quest for faster, cheaper, more efficient and more seamless systems of design and construction.

Susan Buck-Morss, in the introduction to her book *Dreamworld and Catastrophe*, describes the construction of mass utopia as the dream of the twentieth century – the driving ideological force behind both capitalist and socialist modernization. She writes: 'The dream was itself an immense material power that transformed the natural worlds, investing industrially produced objects and built environments with collective, political desire.' Arguing that the dream died with the end of the century, Buck-Morss states that – although commodities continue to be manufactured in greater numbers than ever – the belief that the industrial reshaping of the world can bring about a good society has now vanished.[45]

No object has stood more strongly for the promise of a new, industrialized future than the car. Transformative, modern and magical in their mass production, cars have not only been symbols of the democratic potential of industry, but have also provided a concrete model for the manufacture of all goods. In ideological terms, the car still maintains a touchpoint as magical manufacture, one that reaches far beyond its place as an individual item of consumption. Car factories now sit at the centre of a high-loaded and politicized network of concerns about globalization, automation, unemployment and national power. The attention given to them today reflects not only their economic might, but also the important role they played in shaping the ideas – and systems – of twentieth-century modernity. In 2019, as many of the dominant structures of the twentieth century stand on increasingly shaky ground, it is the incredible weight and significance of this history in our collective consciousness that ensures the car retains a place at the centre of our vision.

ALMOST

Georgina Voss

LIKE

A CAR

Software is an atypical technology,[1] becoming present only in the material space of a machine. What does it mean to 'make' a car, when control is increasingly centralized in the computational means that form the guts of automotive machines? And how does this change the idea or the very nature of what a 'car' is?

In his 1999 essay 'In the Beginning was the Command Line',[2] speculative fiction writer Neal Stephenson marshalled operating systems into an automotive analogy. 'Imagine a crossroads where four competing auto dealerships are situated', he writes. On one corner is Microsoft, which started out selling three-speed bicycles (MS-DOS), which could be fixed when broken. Next door is Apple, offering expensive over-stylized cars whose insides are mysteriously sealed. A newer competitor, Be, Inc. (a now-defunct US computer company), has set up shop full of Batmobiles (their BEOS system). And on the final corner is Linux – not actually a business at all, but a 'bunch of RVs, yurts, teepees, and geodesic domes set up in a field and organized by consensus'. The people there sell tanks – not old Soviet-era juggernauts, but new and better machines, 'jammed with sophisticated technology from one end to the other'.

The passage works in part because of how beautifully it binds the cultural values of each cohort to their products, but also because of the seeming dissonance between computational technologies and automotive engineering. Yet two decades prior to publication of 'In the Beginning', the analogy had already collapsed into the real. In 1977 the first production car to incorporate embedded software rolled off the production line – General Motors' Oldsmobile Toronado, a lovely sleek thing with an almost preternaturally long bonnet and, inside, an electronic control unit (ECU) that managed electronic spark timing [60]. A decade after 'In the Beginning' was published, BMW and Linux were actively trying to develop open software 'for the connected car' through the GENIVI Alliance.[3] And as I write this, autonomous vehicles are being lauded in the media, by governments and through tech PR releases as the next new thing, jammed with sophisticated technology from one end to the other. Yet your average run-of-the-mill car – still on the roads – already contains around one million lines of code.

Several weeks ago, in a tiny vintage shop in Amsterdam, I came across a Bakelite brooch of a car from the 1940s. The shape of this little speckled amber object was instantly recognizable: the bumper, the curvature of the bonnet, the semicircles for wheels. Go to any end-of-year vehicle-design degree show and, despite the slick future-shock exteriors, each of the models on show is undeniably a *car*. Yet throughout, what holds increasing mechanical control (through sensors) and centralized control (through locked-in operating systems) is somehow elided. Automotive failure is still seen as the realm of mechanical breakdown rather than computational control: cast your eye, for example, over the six million software-based vehicle recalls in 2016.[4]

(An aside: although we're talking about land vehicles here, and primarily cars, this isn't solely their affliction – aeroplanes, construction cranes and submarines also increasingly marry a familiar outer face with internal embedded controlling lines of predominantly proprietary code; Britain's fleet of nuclear submarines are equipped with Windows-based command systems.)

Talk of cars rolling off the 'production line' continues to uphold this imaginary, shielding the presence of computational power. When we think of the work that goes into making an automobile, we are thinking about the finished car, front and centre – sprawling supply chains and just-in-time efficiency are not apparent. While the engine capacity of a vehicle can now be adapted remotely through 'over-the-air' interventions,[5] any vehicle CEO worth their salt knows about the fetishistic power of the assembly line. When Elon Musk brings the press into Tesla, Inc., it's to the factory floor of the Fremont premises: a white-walled cathedral, with Musk himself foregrounded in the polo shirt and cap of the engineer on the floor. In the background there is always some fibreglass skeleton frame or a half-undone car being built up into something whole. The focus is on mechanical assembly, with human workers and red or yellow robotic

↑ **60** The General Motors' 1977 Oldsmobile Toronado was the first production car to feature embedded software. It included an electronic control unit that managed electronic spark timing

arms acting – 'working' – together. What work gets done to create the cars' computational guts (all those hundreds of sensors and millions of lines of code, knitting it together on circuit boards, tested to death) is wrapped away, out of view.

Out of sight, we can take the subtle shifts permitted by software infrastructure for granted. Perhaps this isn't really a surprise: after all, functional infrastructure is often invisible, only becoming visible when it breaks.[6] And how better to break systems than through a world-shaking apocalypse?

Come with Me, then, to the Wasteland

You find what you find. You bother to drag back what strikes a chord in your heart. You repurpose it to war. You fetishize it, because it's more important than you are, and then you build it, stick a cup holder on it, and head out into the wasteland.
Colin Gibson, 2016[7]

In the 2015 film *Mad Max: Fury Road* what is in view is a post-apocalyptic landscape studded with things that are – to paraphrase Charles Darwin[8] – *almost* like cars and which have been hauled into being by their owners [61].

From the souped-up motors of the sickly War Boys to the sonic aggressions of the Doof Wagon, the vehicles of *Fury Road* are fucking ridiculous. The philosophy of their design was, the film's production designer Colin Gibson says, to serve four or five different purposes, and it shows in their outrageous forms: echoes here of Jeff VanderMeer's biology-through-a-prism, a Laocoön tangle of bodies and parts. The protagonist Imperator Furiosa's War Rig binds

↑ **61** The fleet of repurposed vehicles in *Mad Max: Fury Road*, directed by George Miller, 2015

a core Tatra T815 drive with a custom front to a widened 1840s Chevrolet Fleetmaster. Other vehicles are equally unsettling: a child's skull with that double row of teeth. Immortan Joe's Gigahorse combines two 1959 Cadillac Coupe de Villes, mounted atop each other in a swaggering slant; the Bullet Farmer's Peacemaker is a Howe & Howe Ripsaw treaded mining rig boshed into a Chrysler Valiant charger, with (of course) part of a Cessna light aircraft strapped on to the front.

The vehicles of *Fury Road* are made from the last automobiles of their kind, from a time when an energy crisis has put paid to the industrial-scale mining, smelting, engineering, welding and mass production that are needed to make cars of any flavour from scratch.

These mongrel, repurposed vehicles are old – part of what the writer and curator Justin McGuirk describes as the film's 'retro future',[9]

where there are 'no new technologies, no new energy sources, and certainly no Tesla home batteries'. (The background presence of petrol everywhere, in a landscape of scarcity where water is hoarded like gold, is something that the film quietly moves past.)

But *Fury Road*'s old machines serve another purpose. After an energy crisis has pulled apart the world, it seems that nothing that is left is digital, and the 'retro-future' specifically locks the world in a pre-computational era. Crankshafts and combustion engines, and thousands and thousands of bullets, are fine and dandy; the presence of a functioning motherboard is not. Any surviving machine whose animus depends on computational systems has probably been left to rust, or been gutted for small parts alone. The vehicles that are lovingly built anew take their bulk from the years when cars were analogue: it is a *1959* Coupe de Ville, a *1940s* Chevy Sedan that persist, adored and worshipped, because they can be repurposed; their action and meaning can be controlled.

Different social groups give different social meanings to things. In the parlance of Science and Technology Studies – aka STS – this is known as 'interpretive flexibility',[10] and it can steer the development of a technology from one terrain into another. In the wasteland, a century-old Chevy becomes something that can be reanimated and defied; the audience can infer that any electric vehicles have long been shredded, scavenged only for parts. By dint of an apocalypse, the psychic conflict between code and car has been shelved: there are no industrial-scale manufacturing or supply chains here. What the War Boys, Vuvalini and Marauders, make is a new hybrid breed: imagined through their own needs and desires, restricted only by the material limitations of the harsh landscape around them.

The Car as Agricultural Machinery

What if constraints around how a vehicle can exist are exerted by its original maker?

If we move from the wasteland to the farm-yard, to American rural life at the start of the twentieth century, we can see the first cars rolling into view. In 1908, via mass assembly, the Ford Model T set the scene as the first affordable automobile, opening up travel to the newly emerging middle classes.

As STS scholars Trevor Pinch and Ronald Kline note, farmers were at first deeply suspicious of these new noisy, dangerous machines, which scared or even killed their animals – 'Devil wagons'.[11] Cars eventually made their way on to farms, not as *cars* per se but more as the idea of what a car could be, steered through the needs and desires of rural farmers and actualized through hard mechanical work. If one were to run a belt over an automobile's spinning wheels, it would transform the vehicle into a stationary power device, which – one Kansas farmer noted – would enable a farmer to 'save money and be in style with any city man'. Adapted as such, cars could be used to run agricultural machinery, including water pumps and wood saws, but could also be put to use on domestic chores, such as powering washing machines or running butter churns. By 1915 one Maine farmer had found so many uses for his car that tax assessors didn't know whether to classify it as a 'pleasure vehicle' or agricultural machinery.

Initially Ford seemed easy with, and even gently delighted by, the malleable identities afforded by farmers to their motors. But manufacturers have their own interpretative frame around their expectations of their products, and the slippery pleasures of early rural use didn't last long. As 'barnyard mechanics' spread, a flurry of new accessory companies emerged, offering up modification kits that could, for example, replace rear wheels and re-form a vehicle into something more tractor-like. In 1916 the Ford Motor Company informed its dealers that converting vehicles into a form not sanctioned by the company would cost them their dealership; and two years later it warned owners that altering their cars in this manner would void their warranties.

In limiting interpretative flexibility, car manufacturers flexed their muscles, closing down the meaning of their cars and pushing vehicle owners into buying their own mass-produced items – tractors, for example, rather than conversion kits or stationary power conversion kits. But to exert this control required companies to wield their power relationships at a distance, via nodes in the supply chain. (Compare this to the imaginary wastelands of *Fury Road*, where auto companies have expired along with the rest of the world – no War Boy expects a tap on the shoulder, admonishing him for tearing apart his 1934 Chevrolet 5 Window Coupe.) Move to the present day, however, and control of the machine itself could be deployed remotely.

A century after Ford dealerships were threatened with sanctions, automotive companies have new ways to limit what a person can do with a tractor, or a car, or something like a car. Over the air or through the wiring, control now has the capacity to be centralized, computationally.

Illusions of Adaptive Automation

So the year is 2019, and you own a vehicle; and, one day, it breaks. Even if it's box-fresh from the factory; even if the marketing emphasizes 'intelligence' and 'agility'; even if you think about the astonishingly complex networks of production and just-in-time flow and orchestration that have brought this enormous mobile computer to you, and you have to lie down for a while. Even then, one day, it breaks, and rather than being able to get under the bonnet and fix it, you find that in this nest of sensors and software you're locked out of your own machine. (Is it even 'your' machine?) One day you find yourself using a USB-to-tractor cable to jack a Windows laptop into your John Deere combine harvester to upload an unauthorized version of Service Advisor.[12]

John Deere is the brand name of Deere & Company, which manufactures agricultural vehicles and other heavy machinery, and which has begun to transform into a software company that runs its technology on tractors – somewhere in the balance between hundreds of software engineers developing enormous complex codebases, run through tens of thousands of hours of testing on 30-ton equipment, and farmers with tight timeframes in which to get their equipment up and running to plant or harvest. Some amount of tinkering is permitted by John Deere, but only up to a point: cross that line and farmers can be sued.

Marketing bumph for these machines abstracts away from the computational: one website for John Deere's Service Advisor shows a man in overalls handing a customer an actual cardboard box with 'Service Kit' printed on the side. There are workarounds, because there are always workarounds. At the policy level, the emerging Right to Repair movement pushes for legislation which permits consumers to repair and modify their – 'their' – electronic devices.[13] A secondary market – an *aftermarket* – has begun to crawl up through the cracks, to craft engine computers and their ilk to be patched into your car. Modern cars are so tremendously complex that an aftermarket unit might need to spoof the presence of the original, so that the non-engine parts work properly – a gearbox is often electronically controlled and may expect, or require, the original unit.

The thing is the same, but it is not. These processes are allied to, but distinct from, what the film-maker Astra Taylor describes as 'fauxtomation' – an obfuscation of the work that goes on alongside apparently autonomous machines, giving the illusion that they're smarter than they actually are. The sleight of hand that pulls the eye away from the computational systems embedded in automotive engineering is not fauxtomation per se – not only because a vanishingly small percentage of current vehicles are fully autonomous, but also because self-driving cars are simply those for which the software already present in so many vehicles today is given primacy.

The locus of power and control in automation has quietly shifted to computational systems over past decades. With this move, the nature of what it means to make, own and mend a car has radically changed, but the *mise en scène* of the factory floor and production line still recalls a time of mechanical control.

Automotive work has always been adaptive: there is no point when the car simply *is*. Despite the controls laid down by the design process and threaded into production, vehicles are used in massively disparate spaces and places, to different and physically gruelling ends. They flicker through diverse identities and meanings. You find what you find. The aftermarket is the space where the adaptive work is done to bring these meanings to fruition. It holds up an angled mirror to the factory floor and shows what is missing from the imaginaries and slick presentations of the automotive industries, but also where seemingly seamless computational command-and-control can have a knife (or new engine part, or chunk of code) jimmied into it. *Here* are the possibilities for doing the work that binds together all parts of the automotive system, dragging it back, repurposing and building it, so that it can head back out into the world, on a different path.

MAKING

Esme Hawes

THE MODERN

CONSUMER

↑ **62** A presentation drawing produced for General Motors' 1954 Motorama show. Orchestras, singers and dancers featured at these annual trade shows, which showcased the latest models

In 1956 General Motors released a promotional film to advertise its annual travelling automobile trade show, Motorama [62]. In the style of a Hollywood musical, *Design for Dreaming* tells the story of a woman awoken by a masked man and whisked away to the New York Motorama. There she flits around the crowded showroom, pushing through swarms of eager consumers clamouring to ogle the latest General Motors models – a Corvette, a Buick, a Pontiac – her mysterious chaperone promising to buy her one of each. After baking a cake in the labour-saving 'Kitchen of Tomorrow' and watching a fashion show of designer ensembles to complement her brand-new cars, the couple jump into the General Motors concept Firebird II and drive off into a futuristic landscape, singing their love of technological progress and consumer choice.

Design for Dreaming is a spectacle of frenzied consumption, an all-singing, all-dancing advertisement for General Motors' latest models. Although excessive in almost every way, the film encapsulates the craze for product styling and new systems of selling that emerged in the early twentieth century. The auto industry's ability to adopt the latest fashions and styles was instrumental not only in defining the look of cars, but also in developing consumer habits that prevail today.

Crafted Luxury

Just 30 years previously there were far fewer options in the car market, which was split between spartan off-the-line mass-produced cars and custom-made luxury brands. Ford's moving assembly line had brought the car within economic reach of many Americans by the mid-1920s and, ironically, in so doing, had dispelled the novelty of the automobile, causing the market to reach saturation point.[1] As assembly-line cars became ever-more common and cheaper to produce, the distinction between those who could afford one and those who could not became diluted.

For the wealthy, this distinction was sought through craftsmanship and bespoke design. A typical example of the era was French patron Suzanne Deutsch de la Meurthe, who in 1922 purchased a Hispano-Suiza H6B chassis, one of the highest-quality chassis on the market, as exhibited at the 1919 *Salon de l'Automobile* [63]. She then sent it to Henri-Labourdette's coach-building workshop to be custom-made in one of the firm's iconic 'skiff-torpedo' bodies. Handcrafted from wood and designed specifically for each chassis,

↓ 63 The 1919 *Salon de l'Automobile*, held at the Grand Palais in Paris, showcased new production and concept cars

Conduire une Peugeot, c'est
"ÊTRE A LA MODE"...

↑ **64** Produced by Peugeot, this brochure emphasizes the link between their cars and fashionability, 1933

this commission would have been a hugely expensive and time-consuming undertaking,[2] but it signified clearly, to her peers and to onlookers from the street, that Madame de la Meurthe was a woman of refined tastes and deep pockets.

Advertisements for car bodies played upon this desire for individuality and craftsmanship, aligning bespoke cars with elegance. One endorsement for a French coach-builder affirmed, 'your elegance requires a suitable casing' and asked, 'if you do not wear off-the-peg clothes ... why choose the same car as everyone else?'[3] Other less-nuanced adverts played on the link between cars and fashionability, such as Peugeot's brochure in which the tagline screamed, 'To drive a Peugeot is ... TO BE IN FASHION.' The car it was selling could barely be seen, but the woman driving it exuded class and glamour [64].

Cars were increasingly synonymous with style. In the same way that clothes communicated status and fashionability, the idea that the car could be yet another accessory in a wardrobe was becoming more widespread. But this notion was far from egalitarian: the polarized car market, with its choice between either mass produced or bespoke, left little room for middle-class consumers to opt for the costly luxury of style.

Choice and Illusion

Under the leadership of Alfred Sloan, General Motors set about plugging the gap in the market, seeking a way to overthrow the monopoly of Ford's reliable and affordable Model T and provide consumers with a different type of product. In 1921 Sloan began to implement a new policy known as 'annual model renewal'. The strategy would offer 'better quality' cars by producing better-looking cars, shifting the focus from function to form.[4] Ultimately, the engineering would stay the same, but the appearance of the car would be styled differently, to give consumers the impression that new equated to superior. Not only would Sloan create more stylish cars, but also 'continuous, eternal change', outdating previous models one stylistic tweak at a time.[5]

General Motors' first attempt to repackage the car was the 1923 Chevrolet Superior. The chassis and engine adopted the same nine-year-old technology used in previous models, but the body was of the newest style, with a lower roof, higher bonnet and more rounded lines. Probably the most noticeable change, however, was the use of colour. Previously cars were painted black or dark grey, as the practicality of these cheap and fast-drying lacquers fitted with the fast-moving assembly line. In the search for new automobile finishes, General Motors collaborated with DuPont to create 'Duco'. This quick-drying, inexpensive and durable lacquer resin could be produced in a kaleidoscope of different colours, and in their first year of production DuPont released 24 shades each of blue, green, brown and orange [65].[6]

The restyled Chevys were thus painted in a range of hues, which boosted sales to a new peak. Colour became an inexpensive way for General Motors to differentiate its brands while adding value to the entire line. Following the success of the brightly painted cars, General Motors hired H. Ledyard Towle in 1925 to design the latest and most complementary colour combinations. Having worked as a camouflage artist in the First World War, devising colour and pattern schemes used on dazzle ships, Towle applied his knowledge of visual deception to product design – if he had the skills to conceal an object, then he could also make that same object conspicuous, through the reverse use of colour. In the same way that he had designed visual schemes for dazzle ships, Towle set about using colour combinations to disguise the basic body shell that was used uniformly across many General Motors models. This made the different models appear fully restyled, while avoiding the huge expense of tooling up and reconfiguring the factory to change a body shape.[7]

Before long, the idea that colour had market value caught on among not only other car manufacturers, but also makers of mass-produced goods in general. Theories about colour and trend forecasting, which had been developed in previous years, suddenly gained popularity. The painter Henry Fitch Taylor, for instance, who had studied colour combinations and published his findings in the *Taylor System of Organized Colour,* expanded his visual library of complementary colours and began marketing it to product designers. Car manufacturers

To be Modern is to be Colorful

↑ **65** Advertisement for DuPont Duco car polishes for its Autumn/Winter colour combinations, published in the *Saturday Evening Post*, 14 September 1929

latched on to tools used by the American fashion industry such as the Textile Colour Card Association, which issued colour forecasts in the form of cards twice a year. These looked towards Paris fashion trends and mimicked couture colours from season to season. After the House of Worth introduced a particular shade of brown in 1931, low-priced cars painted in Duco's imitation 'Worth Brown' sold well for the next two years.[8] As coloured products gained popularity, product designers recognized the value of colour and followed suit, from Kodak's Beau Brownie cameras designed by Walter Dorwin Teague, to Corona typewriters (also coloured with Duco paints) and even General Electric's Monitor Top refrigerators in 1933.

As colour-trend forecasting had proved, Sloan's notion of obsolescence – that style could date products more quickly and reliably than technology – was not innovative to car manufacturers. The fashion industry, with its ever-changing colours, silhouettes and lines, had been practising styled obsolescence for decades. The first fashion

shows, or 'fashion parades', were held twice a year in Paris couture salons in the early 1800s, and by the late 1880s it was standard for couturiers to have biannual seasonal collections, Spring/Summer and Autumn/Winter. Although fashions changed before this, the emergence of the fashion show emphasized the concept of seasonal change, marking out set periods within the year when clothes were acceptably fashionable or when a new wardrobe was required. The similarity between the two industries was not accidental. In Sloan's words, the 'laws of the Paris dressmakers have come to be a factor in the automobile industry – and woe to the company who ignores them'.[9]

Sloan might well have been referring to the rapidly changing fashion trends taking place within motoring culture itself, trends that changed the look of a garment but ultimately retained its functionality. As early as summer 1905 Jeanne Paquin's Parisian salon produced a series of motoring coats: a brown duster coat and hood, a dark-grey ensemble with shoulder capes and billowing sleeves, and a fitted green cloth coat with matching hat [67].

Remarkably different designs followed the next season. Paul Poiret's *Manteau d'Auto* of 1912 repurposed the design of a North African *abaya*. With blue silk raglan sleeves and an asymmetrical button closing, it perfectly encapsulated the oriental influence that was popular in clothing at the time [66]. Each of these coats would serve its purpose, in keeping the wearer warm and dry while driving, but would appear dated when new products became available, so diminishing their stylistic value even while their utility remained unimpaired.

But the problem facing mass-producers was how to meet the escalating demand for stylish goods without undermining economies of scale. During the 1925 *Exposition Internationale des Arts Décoratifs et Industriels Modernes* we begin to see demonstrations of how this could be achieved – and how the decorative-arts industry could evolve in tandem with the modern technology of mass production. Bringing together designers and industrialists to produce aesthetically pleasing, mass-producible goods adapted for urban lifestyle and moderate-income households,[10] the exhibition was a sign that the

← ← **66** Paul Poiret's stylish interpretation of a motoring coat: the *Manteau d'Auto*, 1912

← **67** Three different styles of motoring coat designed by couturier Jeanne Paquin for her summer 1905 line

↑ **68** The first consciously styled factory-produced car, the LaSalle Convertible Coupe, with designer Harley Earl at the wheel, 1927

increased public demand for style was pushing mass-manufacturers to radically rethink their models of production.

Sloan's efforts to keep up with stylistic changes were realized in 1927 with the GM LaSalle [68]. Designed by Harley Earl, it was the first factory-produced car to be consciously 'styled' – in that it sought a consistent and calculated look to appeal to the aesthetic tastes of a particular market. The overall proportions of the LaSalle echoed the grace, elegance and speed of luxury handcrafted cars. It was longer and lower than other production cars and all the sharp corners were rounded off, thus replacing the mechanical look of rectilinear lines with the superficial appearance of unity and craftsmanship. Not only did this new body shape emulate the look of luxury, but it also helped to camouflage the telltale signs of a mass-produced vehicle – the fragmented cluster of dissimilar parts, signs of assemblage and crudely finished functional interior were all concealed by the cleverly designed body.

L'Opéra

LA SALLE — CAR OF THOSE WHO LEAD

Wherever the admired and the notable are gathering, observe the frequency with which a La Salle rolls to the entrance. The famous, the beautiful, the social arbiters —the roster of La Salle ownership is studded with their sparkling names. Sophisticated judges, these, of what is best. In a motor car they demand much—so much that all the beauty and outstanding luxury of the

La Salle would fail to satisfy were it not coupled with the incomparable character of La Salle performance. They are enthused by its Fisher coachwork, its comfort and its impressive, original beauty. They are won even more, by the fact that the La Salle chassis, with its magnificently efficient 90-degree, V-type, 8-cylinder engine is today's highest expression of the automotive art

You may possess a La Salle on the liberal term-payment plan of the
General Motors Acceptance Corporation — the appraisal value of your
used car acceptable as cash — priced from $2495 to $2895 f. o. b. Detroit

CADILLAC MOTOR CAR COMPANY
DETROIT, MICH.   DIVISION OF GENERAL MOTORS CORPORATION   OSHAWA, CAN.

LaSalle

MANUFACTURED · COMPLETELY · BY · THE · CADILLAC · MOTOR · CAR · COMPANY · WITHIN · ITS · OWN · PLANTS.

While the outward appearance of the car seemed brand new, much of the technology and engineering stayed relatively the same. Another classic example occurred a couple of years later with the so-called 'cast iron wonder' – an engine that General Motors introduced to its Chevrolet range in 1929, which remained in subsequent annual models for the next 25 years. While consumers paid higher prices for the new models of Chevy each year, it was essentially the same car on the inside as previously, but dressed up in new clothing.

The LaSalle was an instant and unqualified success. Numerous articles touted it as having the 'look of the finest craftsmanship'[11] and 'all the smartness of the leading foreign cars',[12] with many comparing General Motors' new model with luxury European vehicles, such as the Hispano-Suiza. The comparisons were not gratuitous, for Harley Earl openly admitted that 'I stole a lot of stuff ... [from the] Hispano.'[13]

Even the advertising campaign for the LaSalle brand played upon its similarities to luxury car design. Illustrated by Edward A. Wilson, the series of adverts produced in 1927 and 1928 showed the new brand in various European locations, one of which depicted the car outside the Paris Opera and included the words, 'Wherever the admired and the notable are gathering, observe the frequency with which a La Salle rolls to the entrance'[14] [69]. Hardly subtle, the French angle imbued General Motors' new brand with the same prestige as its authentic counterparts.

In fact such gratuitous borrowing of designs was not a new concept. Much of the ready-to-wear fashion industry had been mimicking luxury products as a way of closing the aesthetic gap in consumption and bringing seemingly 'high-end' goods to a mass market. Haute couturier Madeleine Vionnet produced her acclaimed 'Little Horses' dress for Autumn/Winter 1921; shortly thereafter, the beaded rayon piece was duplicated and sold without the designer's authorization. The German fashion magazine *Die Dame* advertised an exact replica in 1922.[15] Vionnet implemented various initiatives to thwart replicas, such as marking her labels with her thumbprint to authenticate each item [70]. Both authorized and illegal copies of French couture were hugely popular with the American market, and it became commonplace for American ready-to-wear manufacturers and retailers to hire *modistes* to visit the French couture houses at the beginning of the season and steal current trends, which were quickly

← **69** An advertisement for the GM LaSalle, illustrated by Edward A. Wilson, from a series showing the car in various European locations, c.1927. Here it is parked outside the exclusive Palais Garnier Opera House in Paris

↑ **70** 'Pompeian red crêpe dress', by Johanna Marbach, as advertised in the German fashion magazine *Die Dame*, 1922. It is an exact copy of Madeleine Vionnet's couture 'Little Horses' dress

constructed and modified for mass production and distribution to the ever-growing American market.

The overwhelming success of the LaSalle proved the soundness of Sloan's product policy with its emphasis on superficial appearance, so much so that a few months after the car's debut, General Motors established a new staff department to focus entirely on aesthetics, and hired Harley Earl as its head. The Art and Colour Section – the first styling department of an American car manufacturer – set a precedent for the auto industry [71]. The Detroit car corporations scrambled to compete, with Chrysler establishing a styling section in July 1928, and smaller mass-producers such as Hudson setting up divisions the following year. Even Ford's production imitated the General Motors strategy. Competition from General Motors eventually forced Ford to move away from sole production of the Model T in 1927, and the huge complications of shifting gear in assembly-line production were highlighted by the fact that Ford factories were closed for six months for retooling, before the new, more consciously 'styled' Model A could emerge. By 1932 Ford had introduced their v8 engine in 14 different models. Style had become the newest and most important selling feature of the day.

↑ 71 Harley Earl's Art and Colour Section at General Motors, photograph by Michael Furman, c.1927. Its designers pioneered many techniques still in use today, such as clay modelling, full-scale drawings and colour renderings

The Style Wars

With all the major car manufacturers now playing the styling game, competition grew to outrageous new heights as stylists were given more freedom to realize their fantastic visions of the car. Without the regulations that would later be enforced on emissions, safety and fuel, designers were left to tinker with their souped-up creations, concocting a vocabulary of stylistic cues that would outdo not only their competitors, but also their own models. This superficial exercise in attention-grabbing ornamentation ignited a style war that escalated into absurdities.

By the end of the Second World War, and seeking inspiration from other vehicles, Harley Earl turned to the forefront of technology – aeroplanes. In post-war America the visual cues of jet planes were not only imbued with the idea of scientific advancement, but also resonated with the current nationalist ideology, symbolizing America's victory and superiority over past and present enemies. The 1948 Cadillac, inspired by the P-38 Lightning fighter aircraft, sported a pointed nose similar to that of the plane, and chrome scoops along the back, simulating the functional air vents on the aircraft. But perhaps the most aeronautical feature was the little fin on each rear fender.

These two-inch bumps were the first iteration of tail fins. As they gained popularity, car manufacturers jostled for prominence, slavishly imitating General Motors' design feature to create larger and more angular adaptions. Over the course of the 1950s the initially modest ornament morphed into jet-age wings. Ford Styling adorned its smooth, clean bodies of 1949 with elaborate chromed tail fins protruding from the sides. They turned up overseas on models by Holden, Vauxhall, Mercedes-Benz, and even on the Trabant. But Chrysler's stylists boldly took the new trope to extremes. The 1956 Dodge Royal Lancer, one of Virgil Exner's 'Forward Look' cars, sported soaring tail fins and razor-thin roof pillars [72]. And while Earl's rounded and bulbous cars were aesthetically stuck in the age of air travel, Exner's Chryslers captured the lean, angular lines of the jet age and the emerging idiom of the space race. The 'Forward Look' models usurped General Motors' incremental changes and propelled Chrysler's sales forward. General Motors stylists were left stunned and confused, beaten at their own game. But the competition spurred them on to produce even wilder extremities with the 1959 Cadillac range, in which the Eldorado Seville and Biarritz convertibles flaunted 45-inch tail fins that soared over the windscreen.

Each revamp of the tail fin created a stir. Car press photos were taken from the back, and in most showrooms the cars were parked with their tail-finned rears to the street. The feature was added not merely as a symbol of airborne escape, but also as a means of distinction. As the size of the protuberances grew incrementally from year to year, they succeeded in outdating the previous models.

But as quickly as they had come, by the early 1960s tail fins had disappeared and stylists looked towards other quirks to differentiate

their designs. Lights were added, as a cheap way of distinguishing models from one another. By the mid-1950s General Motors' cars, which had managed to get around at night with just three lights, now included up to 14 on the front, rear and sides. On many cars the tail lights were dummies or did not function at all. And only a few years later, in 1960, the new look was achieved by reducing the number of lights. Chrome conical bumpers, nicknamed 'dagmars' after a famously buxom actress of the same name, were tacked on – first on either side, then placed more centrally, and later with the addition of black-rubber tips, until they were omitted altogether. The Buick division created its own styling trademark by introducing chromed portholes in the sides of the front fenders. Marketed as 'ventiports', they oscillated between three and four on each side from

← 72 This advertising campaign for the Dodge Royal Lancer focused on the car's soaring tail fins, 1956

↓ 73 Advertisement for the Chrysler *Dodge La Femme*, a car designed for women, complete with matching handbag, accessories and special interior fabrics, 1955

year to year, were banished from the Buick look in 1958 and 1959, and then returned in a more stylized form on the 1960 line. In the same fashion, panoramic windscreens were added to Chevrolets in 1954 and, for even more choice, could be tinted various colours. Each style swing – justifiable or unfounded – helped in creating obsolescence.

Sloan's vision to create 'a car for every purse and purpose' was being taken to its extremes by other manufacturers. In 1955 Chrysler introduced the Dodge *La Femme*, a car designed specifically for women [73]. It was painted pink – and equipped with a rain hat, coat and umbrella, as well as a pink leather handbag containing make-up and a cigarette case. Both the raingear and handbag could be stowed in special storage bins attached to the back of the front seats. The car was simplified for female drivers with push-button transmission controls – 'the effortless, lady-like way'.[16] This was not the first time a car had been consciously styled for women, for the previous year Chrysler had unveiled its 'His and Hers' concept cars: *La Comtesse* and *Le Comte*. Both attempts were met with disinterest, mostly because women found the crude attempts to distract their attention rather patronising, as, after all, the cars appealed to a classic male ideal of femininity rather than how the 1950s woman actually saw herself. General Motors changed tack in the mid-1950s, hiring a group of all-female industrial designers, believing that they could make cars that would be attractive to women. Dubbed the 'Damsels of Design', they were tasked with bringing 'a woman's touch' to the colours and fabrics of the upholstery, as well as with creating gimmicky accessories and labour-saving conveniences for the family car.[17]

While the majority of these changes were relatively minor, often involving chrome accessories designed to trick the eye into thinking

now for the first time anywhere, a car glamorously, *Personally Yours*

Never a car more distinctively feminine than *La Femme* ... first fine car created *exclusively* for women! In this superbly designed car, Dodge brings together luxurious, delicately-toned interiors and ultra-fashionable appointments ... every sophisticated touch your heart could desire! Here is, truly, the ultimate in fine motoring.

that the body had been reshaped, each addition was unveiled as though it was the latest innovation. Manufacturers relied on the drama of marketing to ceremoniously reveal each new model, creating brochures, promotional films and advertising campaigns, accompanied by celebrity endorsements. Most spectacular of all were General Motors' Motorama shows, during which *Design for Dreaming*, mentioned at the beginning of this chapter, was first promoted. These full-scale entertainment extravaganzas pandered to society's need for diversion and escapism after the deprivations of the Depression and the war. The shows drew from Hollywood film sets, with top performers, original songs, choreography and showgirls [74]. But unlike the cinema, where the escapism and spectacle evaporated as soon as the audience left the theatre, the marvels in these shows materialized as tangible commodities. For once, General Motors was not the trendsetter; Paris had hosted the *Salon de l'Automobile* since 1898, running car trade shows in the Grand Palais that closely resembled the immensity and marvel of universal expositions. They undertook feats such as illuminating the entire glass dome of the building with 200,000 electric lights, attracting unprecedented numbers of visitors and creating 'a fairytale spectacle' to flaunt the models on display.[18] These elaborate exhibitions were constructed with the sole purpose of enhancing the product and whetting the public's appetite.

↑ 74 The unveiling ceremony of the latest concept cars at the General Motors' Motorama, 1956

The Backlash

As stylists strove for novelty in their designs, producing models that were even longer, lower and more extreme, the cars' performance and functionality suffered. The July 1959 *Wall Street Journal* reported on the General Motors annual meeting in which shareholders openly lamented the difficulties they were experiencing with the regular model changes. Some complained that they could not sit in their cars with their hats on, and had taken to wearing bicycle clips to stop their clothes trailing on the floor; others told of bruised knees and heads as they tried to clamber into the lowered driving seat. Most troubling, perhaps, were the numerous criticisms that the lower cars obscured the driver's view of the road ahead, as sight lines were now more than nine inches lower than they were in post-war cars.[19]

Stylistic changes were not only causing minor annoyances. Ralph Nader claimed in *Unsafe at Any Speed* that extended tail fins were among the myriad murderous calamities that carmakers had forced upon the public. He mentioned three fatalities, including nine-year-old Peggy Swan, who had to undergo a tracheotomy after riding her bike into the point of a tail fin in September 1963.[20]

Emphasis on styling led designers and consumers alike further and further away from actual variations that improved the product's function. The designer J. Gordon Lippincott bluntly summed up the frivolity of this practice, stating that contemporary styling 'loses its glamour only one notch slower than a streetwalker at dawning'.[21] In 1960 Vance Packard's *The Waste Makers* had already exposed planned obsolescence being practised by car companies, arguing that such frequent shifts in design not only degraded the quality of products and diminished the planet's natural resources, but were also actively converting the consumer into a 'wasteful, debt-ridden, permanently discontented individual'.[22]

Meanwhile, stylists made attempts to justify these trends for sound functional reasons: the low centre of gravity acted as a great aid in cornering; lower cars were faster and more powerful; and the addition of jutting tail fins could stabilize a moving car in cross-winds. Few designers in Detroit could provide much evidence to support these claims, and none of these add-ons seemed to alter the car's performance.

Other car manufacturers actively scorned the annual model change. Volkswagen's advertising campaign by Doyle Dane Bernbach went against the status quo of car adverts by celebrating the no-frills functionality of the vw Beetle in a series of bare-boned, monochrome adverts. The Volkswagen 'Theory of Evolution' showed the Beetle unchanged, year after year [75]. Deliberately citing the style features employed by the Detroit car manufacturers, it asked: 'Can you spot the Volkswagen with the fins? Or the one that's bigger? Or smaller? Or the one with the fancy chrome work? ... You can't?' In another, a Beetle was spotlit to mimic the presentation of General Motors' annual models at the Motorama show, with the title 'The '51 '52 '53 '54 '55 '56 '57 '58 '59 '60 '61 Volkswagen' – a jibe at the absurdity of obsolete

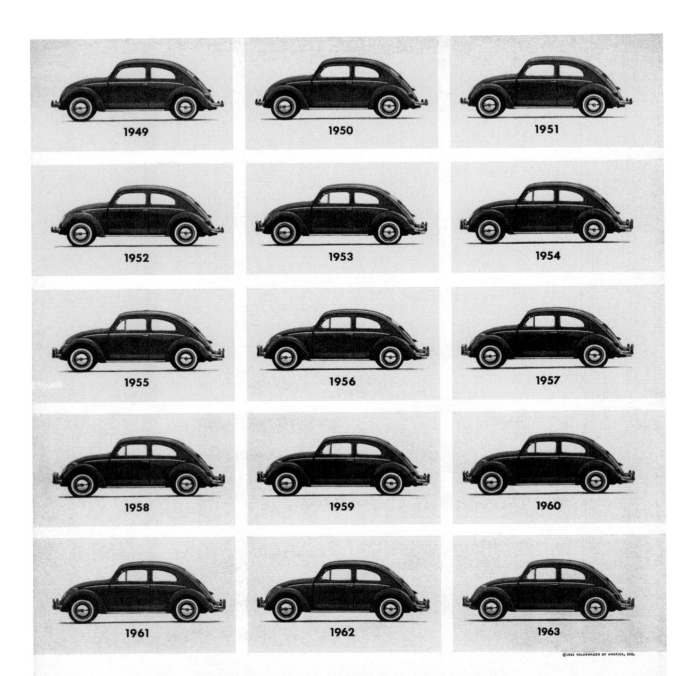

The Volkswagen Theory of Evolution.

Can you spot the Volkswagen with the fins? Or the one that's bigger? Or smaller? Or the one with the fancy chrome work?

You can't?

The reason you can't see any revolutionary design changes on our car is simple: there aren't any.

Now, can you spot the Volkswagen with the synchromesh first gear? Or the one with the more efficient heater? How about the one with the anti-sway bar? Or the more powerful engine?

You can't?

The reason you can't see most of our evolutionary changes is because we've made them deep down inside the car.

And that's our theory: never change the VW for the sake of change, only to make it better.

That's what keeps our car ahead of its time. And never out of style.

Even if you aren't driving the most evolved VW of all.

Our '63.

← 75 Advertisement from a series produced by Doyle Dane Bernach for Volkswagen using pared-down imagery and witty text to emphasize the reliability and functionality of the VW Beetle, 1963

↑ 76 Ant Farm's *Cadillac Ranch* was a staged installation in which 12 Cadillacs were buried bonnet down in Amarillo, Texas, 1974

styling. Hinging their campaign on the theory 'never change the VW for the sake of change, only to make it better', the car gained cachet for being small (economical) and unchanging (reliable).

The art practice Ant Farm joined the critique with *Cadillac Ranch* in 1974 [76]. Staging the performance along Route 66 in Amarillo, Texas, it bought 10 Cadillacs, each representing a step in the evolution of the tail fin, then buried them bonnet-down in chronological order with their boots sticking out. Not only was *Cadillac Ranch* envisioned as a drive-by experience for unassuming road-trippers, but also visitors were encouraged to participate in a free-for-all graffiti spree – marking the car bodies with whatever they brought with them. To counterculture rebels, this was an irresistible opportunity to deface a consumerist icon, while for others the act was sacrilege. This monument to consumerism probably wouldn't have been accomplished without the artists indulging in the practice they were critiquing. When buying the 10 Cadillacs, one member commented that the cars were 'painful to bury'. But in contrast, after being overcharged for a 1949 model, it was suggested that they smash the front end with a sledgehammer. This they did, and filmed it, in front of the bewildered seller as he winced in agony.[23]

The rebellion against mass consumption and the rejection of standardization continued, with the emergence of unique car sub-cultures around the world. Groups created their own visual language for cars, turning them into icons of personal expression and ultimately subverting the passivity of the consumers at the bottom, who await deliverance by the experts at the top. American hot-rodders and Swedish raggare stripped Detroit's chromed giants of their superficial body parts and then modified the mechanics to make them more powerful [78]. Later, Japanese bōsōzoku pushed motorcycle customization to its extremes [77], and Los Angeles lowriders adapted hydraulics and designed extravagant custom paint jobs [79].[24] The car, for these groups, was not an object of passive consumption, but of active subversion.

By 1950 the Detroit car manufacturers were spending more than US$1bn per year to put a new dress on their cars, which added an extra $200 to the cost per vehicle. As the price of a new car was around half of an average family's salary, the rise in price was not insignificant.[25] In the same decade the increasing availability of short-term credit and instalment plans encouraged immediate material gains. The 1950s saw private debt in the USA more than double, from $104.8bn to $263.3bn, with more than two-thirds of all cars being bought on an instalment plan or credit.[26]

Nevertheless, the financial cost of the annual model change has done little to deter customers. The advent of styling and selling techniques in the auto industry trickled down, to be used to manufacture and sell other products. This 'out with the old, in with the new' mentality has created a culture of accelerated consumption. Discussing the abundance of consumer choice in 1960, the senior editor of *Sales Management* magazine speculated that 'If we Americans are to buy and consume everything that automated manufacture, sock-o selling and all-out advertising can thrust upon us ... the only sure way to meet all these demands may be to create a new brand of super customers.'[27] In the same way that the car industry adapted to meet demand, so consumer habits evolved to embrace the influx of products. Today we are perhaps those 'super customers', surrounded by unprecedented amounts of consumer choice. From the latest iPhone to generational colour trends and fast fashion's 50-plus 'micro-seasons' a year, Sloan's 'annual model change' policy still very much prevails in the products that we buy and in our attitudes towards consumption.

→↘ **77 & 78** Around the world, automotive subcultures emerged in rebellion to the homogenization of consumer culture. Pushing against typical car design, subcultures reinvented vehicles for their own purposes. Bōsōzoku in Japan customized motorcycles to include over-sized fairings and lifted handle bars, while Swedish raggare adopted retro American cars

→ **79** Lowrider subcultures based in Los Angeles adapt hydraulics systems and adorn cars with extravagant paint jobs

OH LORD, WON'T YOU BUY ME A MERCEDES-BENZ

Johanna Agerman Ross

In the opening line of one of Janis Joplin's most famous songs she invokes enlightenment through the purchase of a luxury German automobile. In her characteristically raspy voice the American singer-songwriter pokes fun at a society obsessed with buying things as a way of building identity and self-worth. While our attitude to consumption has altered very little in the 50 years since the song was recorded,[1] the symbols of that consumption have transformed drastically.

The car is no longer the powerful identity-builder it once was. Our vehicles are moving from individual possessions to service models shared by many. Simultaneously, as self-driving cars are entering the market, the expectations of a car's performance is also changing significantly. If the car is no longer an expression of the individual owner's desires, but rather a commodity shared by larger communities of consumers, how will that change the look of the car itself? By delving into press releases from car manufacturers, and into reports from market researchers and news agencies, a clearer picture starts to coalesce of *some* of those possible futures.

From Custom-made to Purpose-built

In three core regions – China, Europe, and the United States – the shared-mobility market was nearly $54 billion in 2016, and it should continue to experience impressive annual growth rates in the future.
Report from McKinsey, 2017[2]

Currently there are 3.5 million Uber users,[3] 185,000 Zipcar members[4] and 2.6 million cars registered in London alone.[5] As the scale of these numbers is altering year on year, with fewer registered cars and a higher number of users of shared mobility schemes, it is clear this will significantly affect how we interact with cars and the brands that produce them.

Shared mobility is the collective term for e-hailing apps, such as US-founded Uber or Chinese Didi, as well as for car-sharing businesses – for example, German companies Car2Go and DriveNow, American Zipcar and a new electric-only car-sharing service announced by Didi Chuxing in early 2018.[6] Currently all these companies operate similar business models, where users take out membership in order to access the service, and are charged either by the distance or time it takes to travel. However, they all operate different models in terms of the cars they use. While e-hailing companies typically treat their drivers as independent 'partners' using their own vehicles to carry out the service, car-sharing companies own fleets of cars. Zipcar, for instance (independently founded in 2000), has a fleet of differently branded cars, negotiated with a variety of partners. Car2Go, on the other hand (founded by Daimler AG in Germany in 2005), uses cars only from its own subsidiaries – Smart and Mercedes-Benz. DriveNow (founded by BMW in 2011) also regards the car-sharing service as an extension of its brand-building and uses only Minis and BMWs. Meanwhile, Chinese Didi Chuxing has set up an agreement with 12 different car manufacturers to provide its fleet of solely electric cars. With the advent of these new types of mobility companies, the customer is no longer an individual, but is increasingly large-scale businesses.

The desirability of the car has always been steeped in its materiality. Ever since General Motors introduced the first mass-produced car to be offered in a colour other than black – in this case the Oakland in eggshell blue in the early 1920s[7] – new styles and functions have led the way in marketing for car brands. A century later, viewed through the lens of new mobility models, the mainstream car seems heavily over-specced, packed with features and styling options to make it stand out to the individual consumer. Consider, for example, the new Fiat 500 Collezione, which on its website pushes the car as a fashion accessory: 'Driving it is like wearing it – rock the style and colour of the Fiat 500 Collezione' and 'The rich autumn hues and sensations are reflected in the colour palette, such as the bi-colour Brunello, which combines deep burgundy with dark grey.'[8]

However, those same features are unlikely to persuade the new mobility brands to invest. For the shared-mobility market, a strategy would instead be to strip the car back to a minimum of parts and styling options. McKinsey reports:

> In the United States, a typical vehicle might cost nearly $24,000 because it represents a compromise developed to appeal to the broadest spectrum of consumers in target segments. A purpose-built vehicle, on the other hand, could feature lower levels of complexity; less powerful engines; simpler, easier-to-clean interiors; less complicated assembly processes; and lower distribution costs. Such a car could cost almost 25 percent less than a typical vehicle [for the individual consumer market].[9]

Didi seemed to be the first to experiment with the model of purpose-built vehicles when, ahead of the Beijing Auto Show in 2018, it announced an alliance with 31 auto-industry partners to develop 'unified standards for the design and manufacturing of new energy vehicles, development of intelligent driving technologies, and planning of charging facilities' across its shared-mobility business.[10]

Some researchers imply that the move towards specced-down cars and the increasing reliance on new car-sharing models are incremental steps towards gaining consumer acceptance for autonomous vehicles, as they are no longer consumed as expressions of self or as identity-builders. When car usage is reduced to simply getting you to places, the car's style or make is of little importance and we cease to regard them as personal expressions of taste; instead, convenience becomes the deciding factor for our choices. 'Car sharing is really the mode that allows us to make the transition to automation,' said Adam Cohen, a researcher from the Transportation Sustainability Research Center at the University of California, Berkeley, in an interview with *Forbes*. 'We can't go from zero to sixty. We need something that allows people to transition out of private vehicles. We need something that allows developers to reduce their parking before we get to a zero-parking scenario, if that is where we're going.'[11] It's somewhere in this hinterland of self-driving

experimentation by the likes of Uber, Google and Didi that the future of the car as identity-builder really lies. Through automation, our expectations of the car's functions will alter completely, and getting from A to B will be just one of many services that the car can deliver.

With the development of General Motors' Art and Colour Section under Harley Earl in the 1920s and '30s, the car's identity was largely exteriorized. And with the arrival of the so-called 'Damsels of Design' – the all-female design department, founded in the 1950s with a focus on designing interiors – the look of the inside became as important as the outside. Through paint jobs, chrome details, custom-upholstery and accessories, the car became a consumer product that packaged the driver's identity into one neat, streamlined envelope, designed inside and out. What is set to happen 100 years on is an increased focus on the largely untapped potential of the car's interior functions and the endless possibilities of reshaping the use of that space in the light of automation.

From Exterior to Interior

Vehicles operating in an autonomous (e.g., driverless) mode can relieve occupants, especially the driver, from some driving-related responsibilities. When operating in an autonomous mode, the vehicle can navigate to various locations using on-board sensors, allowing the vehicle to travel with minimal human interaction or in some cases without any passengers. Therefore, autonomous vehicles give passengers, especially the person who would otherwise be driving the vehicle, the opportunity to do other things while travelling. Instead of concentrating on numerous driving-related responsibilities, the driver may be free to watch movies or other media content, converse with other passengers, read, etc., while riding in an autonomous vehicle.
Patent US9272708B2, Ford Global Technologies, 2016

In March 2016 a subsidiary of Ford Motors applied for a patent called the Autonomous

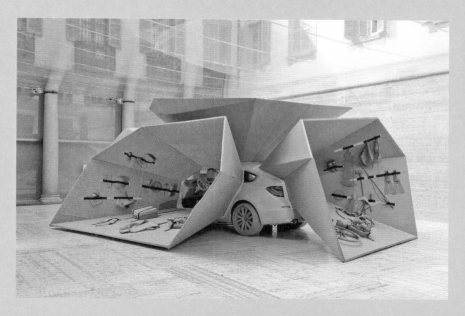

↑ **80** Designers Patricia Urquiola and Giulio Ridolfo collaborated with BMW to create 'The Dwelling Lab', transforming the interior of a BMW 5 Series Gran Turismo to include an array of innovative products and interior details, 2010

Vehicle Entertainment System. The system darkens a car's windows, lowers a screen to block the front windshield and shows movies to passengers.[12] Of course, with driverless cars the architecture and use of the car are free to change completely, as few of the car's design features that we now take for granted would be needed any more. 'We need very modern designers because all cars right now are more or less the same on the inside,' said the Spanish designer Patricia Urquiola in a 2018 roundtable on self-driving cars.[13] Urquiola's work includes creative direction for furniture brands and she frequently designs the luxury interiors of hotels and stores, but in recent years she has also considered the many possibilities of a car's interior space. In 2010 she presented 'The Dwelling Lab' with BMW and fellow designer Giulio Ridolfo, for the presentation of the BMW 5 Series Gran Turismo [80]. In the installation the exterior of the car was concealed by four cone-like elements, upholstered in buff-coloured textiles, protruding out of the car's open doors. Visitors were invited to explore a soft and embracing car interior, containing objects more usually seen in your home, such

as clothes, crockery, gaming devices, a dog bed and a skateboard. 'The Dwelling Lab allows you for the first time in history to see the interior of the car before you see the exterior. It highlights the growing importance of a car's interior,' said Adrian van Hooydonk, BMW Group Design Director at the time.[14]

BMW's venture into art installations of car interiors somewhat proves Cohen's point of needing 'something that allows people to transition' from cars as we know them, to cars as they might be in 10 or 20 years from now. Design is of course a useful tool in this process, expanding the possibilities available by creating fictitious scenarios for us all to explore. However, when it comes to the future of mobility, it's not just car brands that are free to explore. The world's largest furniture retailer, Ikea, recently presented seven concepts for activities that passengers could undertake while in transit in self-driving vehicles, through its 'future living lab', Space 10 [81]. The minibus-sized containers, in a range of friendly pastel colours, housed cafés, offices, doctors' surgeries, farm shops, gaming facilities and lifestyle stores. In a press statement

↑ 81 'Space on Wheels' is a project by Ikea's innovation lab, SPACE10, exploring the potential for self-driving vehicles to take on new uses and support daily activities. This space envisions a mobile clinic, 2018

about the concept launch, Ikea's concept innovation manager, Göran Nilsson, stated:

> We don't have ambitions of manufacturing cars, but in a future where people no longer have to worry about driving, vehicle interiors can expand to a point where we no longer are designing cars, but rather small spaces. Then it's suddenly an area where we have a lot of experience, but also an area where we would like to engage for new insights.[15]

From the Car to the City

More people than ever are living within a shrinking footprint. With MINI LIVING, MINI is addressing these developments and creating the first co-living project in China. MINI is working with Chinese project developer Nova Property Investment Co. to transform an unused industrial complex in the Jing'An district of Shanghai into a multi-layered co-living initiative made up of apartments, working spaces and cultural/leisure offerings.
Announcement on Mini website, 2018[16]

On the morning of 10 May 2018 four Greenpeace activists pulled up in front of the V&A Sackler courtyard in a VW Golf TDI, with the banner 'The Future Doesn't Start Here'. Slowly and methodically they started taking apart the car's diesel engine and exhaust system, piece by piece, laying it out in a pattern on the light porcelain paving. It was the day of the opening of the new V&A show *The Future Starts Here*, with the lead sponsor being VW. The Greenpeace campaign was calling on VW to abandon all diesel cars and to invest in electric instead. The campaign highlighted another function of the styling of cars: that of concealment, of hiding the polluting core of the vehicle under a stylized and brightly coloured, polished shell. As the recent report from the Intergovernmental Panel on Climate Change made clear, hundreds of millions of lives are at stake, should the world warm by more than 1.5 degrees Celsius, unlike the two degrees that was agreed at the 2016 Paris Climate Accord. With fossil fuels accounting for a substantial part of the earth's warming, electric cars are already a viable alternative, and more and more car brands are releasing their own versions of electricity-powered engines.

↑ 82 The Mini Living building initiative, launched by the car company in 2016, showcases ideas for how to maximize space and quality of life in compact urban environments

The number of electric cars registered in the UK has surged in the last decade from around 3,500 in 2013 to almost 119,000 in 2017.[17] However, in the light of new mobility models and the arrival of automated vehicles, some car manufacturers are already starting to contemplate the total obsolescence of the car as we know it, using their brand position and market reach for new ventures instead. In 2016 Mini presented the Mini Living initiative at the Milan Furniture Fair, with the aim 'of devising creative architectural solutions for the urban lifestyles of the future' [82]. The first habitable Mini Living compound is already under way in Shanghai, in a post-industrial complex redesigned by London-based architects Universal Design Studio. In the renders of the project, cars – except for one Mini – are curiously absent. Instead the streets are inhabited by people and trees, creating tranquil surroundings where it seems that most amenities, including your place of work, are in the same place as you live. In these soon-to-be-real visualizations, it is not the car that reflects our lifestyle; instead the car has been subsumed by lifestyle.

After a century of the car being sold as a reflection or extension of ourselves, through make, model and styling, we are now being asked to consider a society without the car. Some would argue that this is a more urgent thought experiment than others, but the wheels are well and truly in motion towards a future where the car's physicality will be drastically redefined. In the 2018 film *Rams*, about the German designer Dieter Rams who achieved fame for his minimally designed products for Braun in the 1960s, the product designer offers a poignant reflection on the future of the car: 'It starts with the design of a landscape – not a machine!'[18]

SHAPING SPACE

Lizzie Bisley

THE RACE TO EXTRACTION

UN QUARTIER EMBROUILLÉ.

In the afternoon of the 12th of December 1955, following a small accident of which the cause remains unknown, a violent electric storm, a tornado ... was unleashed on western Europe bringing profound disruption to everyday life.[1]

So begins Albert Robida's 1892 novel *La vie électrique* (The Electric Life). Robida tells the story, set in a future France of the 1950s, of a world revolutionized by electricity – describing a planet fitted out with flying cars [83], illuminated glass skyscrapers, international televisual communication and intergalactic telephones. In Robida's universe, the power of electricity has been harnessed so completely that even nature is now under its thrall – the world's climate has been engineered to create an entire planet of temperate, fecund conditions, with electric currents used to shift storms away from cooler continents to hotter ones, and glacial seas circulated globally. As the opening tornado suggests, however, all is not well in Robida's fictional world, and the electric current reveals increasingly dark and dystopian sides as the story unfolds.

La vie électrique was the third of Robida's novels, following two similarly themed (although more optimistic) works published in 1882 and 1887. The books were almost exactly contemporary with the advent of the car, and offer a tantalizing glimpse into the technological imaginary of the late nineteenth century. Painting a dizzyingly extreme vision of the collapse of distance that was embodied in contemporary technologies such as the telegraph, the world created in *La vie électrique* is governed by speed, movement and human ingenuity, all channelled through the fearsome conduit of electricity.

As a genre, science fiction often imagines future consequences for the machines and inventions of its day. With cities and landscapes governed by fleets of individual flying vehicles, *La vie électrique* creates a picture of the role and future that Robida's generation saw for the car: universal and electric-powered, offering boundless mobility and freedom of movement. In its vision of a planet completely reshaped by motive power, the novel also hints at the automobile's most lasting and terrible footprint – although many of Robida's generation would have been surprised to find gasoline, and not electricity, as the fuel that has (until now) shaped the car's environmental legacy.

It is hard to imagine a single technology that has made more of an imprint on the globe than the car. The automobile's long-term impact has in part been formed by exactly those images set out in Robida's fictional world: the idea of a vast space, ready to be explored and exploited by those humans who are in control of the machine.

'Who knows what miracles electricity will bring us in this area?'[2]

The earliest automobiles emerged in the 1880s from a soup of other technologies. In construction and form, they were particularly strongly influenced by both the carriage and the bicycle, borrowing and modifying designs for wheels, tyres, suspension systems and bodies.[3] This process of adaptation was accompanied by a strong sense, in early writings about the automobile, of a vehicle whose final form had yet to be defined. Technical magazines reported in detail on the latest cars developed in both Europe and the USA, highlighting innovative systems and speculating on the future of different designs. The focus was on establishing which was the most promising new model, with lengthy descriptions of cars that had

↑ **84** General Electric promotional photograph of a woman charging an electric car in her garage, c.1900

finished long-distance drives, competed in races or been shown in public exhibitions.

The question of the future of the car was almost always centred on the issue of motive power. By the 1880s three dominant types of motor had been developed – steam, gasoline and electric – and the relative benefits of each were discussed at length by makers, lobbyists and the press, with much speculation about which would become standard (as well as whether alternate forms of fuel would be developed instead). The French engineer Gérard Lavergne summed up the nature of the debate in his 1900 book *Manuel théorique et pratique de l'automobile sur route*, in which he described in detail all the steam, gasoline and electric motors in production, offering an appraisal of each type. Steam motors were generally deemed by Lavergne too heavy for touristic or everyday use, but perfect for heavy loads; while gasoline, praised for its speed and power, was also referred to as smelly, inflammable, noisy, prone to cutting out and only suited to long-distance touring. It was electric power that Lavergne described as the automobile motor '*par excellence*': highly reliable, easy to drive, quiet and perfect for city use.[4]

Lavergne's arguments were mirrored much more widely, on both sides of the Atlantic, in a strongly held belief in the potential of electricity as the fuel of the future. Although electric batteries did not yet have the range of the petrol engine, many believed that the technology would soon catch up, and that its other advantages far outweighed this (temporary) blip. As *Scientific American* stated in 1899:

> The storage electric motor is clean, silent, free from vibrations, thoroughly reliable, easy of control, and produces no dirt or odor ... The greatest demand for an efficient automobile comes, not from people who wish to take long tours through the country, or whose business calls them to any considerable distance from an electric charging station, but from surgeons, expressmen, and those private citizens who wish to keep a carriage.[5]

The equivalence between electric motors and convenience extended beyond the surgeon and the expressman into a strong campaign to make the electric car the vehicle of everyday use. Interestingly, this was done in part by trying to capture a large potential market of women drivers, with manufacturers in the USA directly targeting women by ascribing the electric car with supposedly 'feminine' characteristics: quiet, smooth, domestic and easy to control.[6] General Electric launched a series of advertisements to this end in the early 1900s, in which women were photographed recharging their cars from rectifiers installed in their own garages [84]. As one advert stated: 'any woman can charge her own electric with a G-E Rectifier ... there are no tiresome trips to a public garage, no waiting – the car is always at home, ready when you are'.[7]

In contrast to this picture of the middle-class woman safely recharging her car for a quiet drive through city suburbs, the image that developed around the early gasoline motorist was one of speed, recklessness and long distance. The petrol car, more likely to break down, noisier and able to travel further distances without the need for a charging point, quickly became associated with adventure, risk and, depending on who you were speaking to, danger and disruption. As described in the German magazine *März*, it 'invokes quiet jealousy in the excluded proletarians, anger with the rural population suffering damage, and with the elite it feeds the old privileged instincts of arrogance and callousness'.[8]

Gijs Mom has argued that it was precisely this association between gasoline and adventure that led to the eventual elimination of steam and electric cars in favour of the petrol engine. Noting that early electric cars were in fact more reliable and a better-performing technology, Mom convincingly describes the triumph of the combustion engine as an example of the power of cultural attitudes to determine a technological shift.[9] By as early as 1900 petrol cars had already started to pull ahead, with drivers embracing an ideal of the complicated fast engine that could travel off-road under the hand of a skilled, sporting driver.

↓ 85 Still from the film *Le Raid Paris-Monte Carlo en deux heures en automobile* (1905), directed by Georges Méliès. King Leopold is seen here as he prepares to leave Paris

'Before starting out for a ride your first duty is to see that the petrol-tank is full'[10]

In 1905 the film *Le Raid Paris–Monte Carlo en deux heures* was released to great international success [85]. Made by the French director Georges Méliès, it follows the Belgian King Leopold II on a fictional drive from Paris to Monte Carlo. The King is the perfect caricature of Mom's reckless, aristocratic driver, mocked for his frequent crashes, speed and lack of control. Some of the most ridiculous moments in the film occur in the search for petrol–the hapless Leopold stops in a small town to fill up, managing to run over several villagers while waiting for a can of petrol to be brought out and poured into his engine.[11]

As the image of Leopold filling up at a small town garage makes clear, one of the massive advantages of gasoline over electric was the fact that petrol cars did not need a substantial refuelling infrastructure in order to travel long distances. In a petrol car, additional fuel could be carried or picked up – with a minimum of equipment – from any number of places en route [86]. This ability to refuel and keep driving, while useful for the sporting driver, also meant that the car could be adapted as a highly efficient tool of exploration and colonial power.

Much early writing about the advantages of car technology focused on its potential for flexible, fast imperial governance. Huge numbers of articles were published about car services that had been established to link up small outposts across European empires, noting the help these could give in increasing control and communication over areas far from metropolitan centres.[12] Colonial outposts of the British Royal Automobile Club were quickly established, and a group of 1920s' despatches sent to the Commonwealth Office by the Royal East African Automobile Association gives a sense of the active role that gasoline cars played in establishing and maintaining imperial networks. Filled with photographs, and written on letterheaded paper illustrated with drawings of African animals in the wilderness, the despatches report on the building of new roads, together with observations made by members on tours through rural areas, and driving conditions in different countries around the region.[13]

As the drawings and photographs of African wilderness suggest, the car had a highly symbolic – as well as an important active – role

↑ 86 The explorer and car enthusiast Michael Terry (1899–1981) led a number of expeditions around Australia in the 1920s and '30s. This photograph, which shows the supply of Shell petrol that was used to keep the expedition going, was taken in Central Australia, c.1930

→ 87 Map showing the route of the *Croisière noire*, 1925

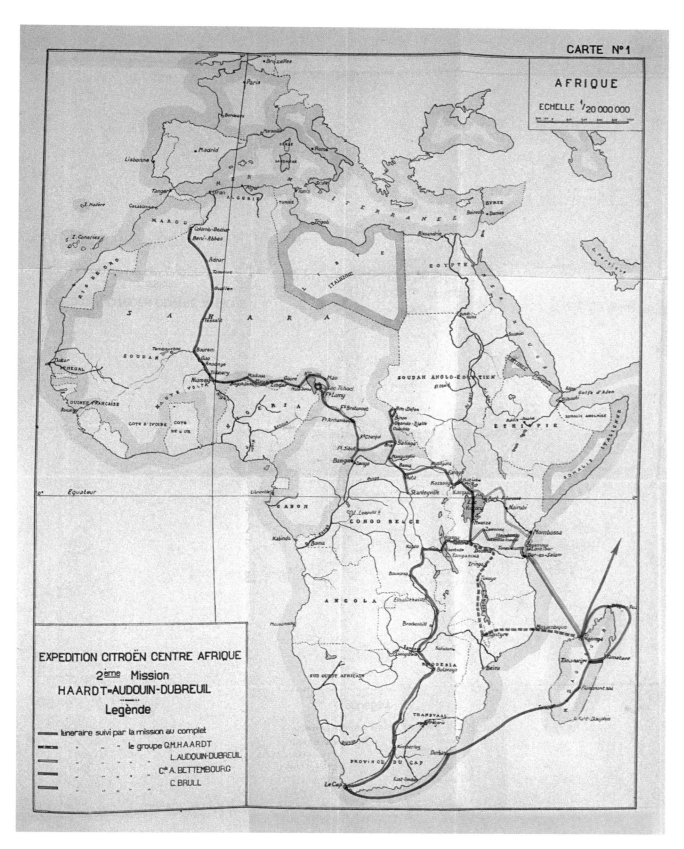

to play in colonial power. These varying levels of significance were expressed most prominently, in the interwar years, in the Citroën company's much-publicized 'Croisière noire'. Launched in 1924, the *Croisière noire* (literally, 'black crossing') was a journey from Colomb-Béchar, in southern Algeria, through West and Equatorial Guinea and south to Madagascar [87]. Intended to link up France's African colonies, the journey was born from a government-backed plan to establish a regular tourist route through the African continent. André Citroën believed that cars could provide a lighter, more efficient and wide-ranging alternative to railroads, and his company designed a series of special off-road vehicles for the challenge – their rear wheels driving on a continuous rubber belt, to allow them to move easily over sand and other difficult terrain.[14]

Although the regular tourist route was never established, the *Croisière noire* garnered a vast amount of publicity. The expedition was documented extensively through film, books, paintings and the display of artefacts collected along the route. In this way Citroën created a very compelling myth of advanced French technology, travelling into the supposedly remote regions of the world [88].

↑ **88** Here the *Croisière noire* cars are in a Congolese village, in one of a huge number of photographs taken to record the expedition, 19 March 1925

During the journey the travellers' progress was reported through newspapers and radio broadcasts; and the film, when released in 1926, drew huge audiences.[15]

Georgine Clarsen has written about the use of cars and bicycles in rural Australia at the turn of the twentieth century, noting that cars were not only useful as a means of communication and supply, but also served to emphasize the supposed technological superiority (and, therefore, the right to rule) of the white settler population over indigenous peoples.[16] In the *Croisière noire*, petrol cars had an important role to play in the shifting political and economic conditions of the interwar period. In the wake of the First World War, at a time when European powers were seeking to reinforce their place on the world stage, Citroën's expedition created a compelling picture of French technological sophistication and imperial might. The petrol-fuelled car, travelling into unknown wilderness, was incorporated into the heroic myths of empire: a new way of bringing modern, progressive 'civilization' to the non-European world.

'a triumphal march for petroleum'[17]

Citroën's African crossing took place at a moment when the power behind the petrol-driven car was becoming an increasing concern for industrial nations. In the 1920s petrol was still a relatively new source of fuel. The first commercial oil operation was established at Balakhani (now in Azerbaijan) in 1837, with the earliest North American oil strikes in Ontario and Pennsylvania in the 1850s. Crude oil was initially refined into kerosene and used as fuel for lights, and it was not until the invention of combustion engines in the 1880s and '90s that gasoline began to have a use.[18]

In the early twentieth century, in step with the growing dominance of the petrol car, gasoline became an increasingly valuable resource. The importance of having control of a national supply was recognized almost immediately, with the French government, for example, sponsoring efforts in 1901 to develop a car powered by alcohol: 'for the praiseworthy purpose of using alcohol as a motive agent instead of foreign petroleum'.[19] While the USA had large domestic stocks of crude oil [89], most European countries had none. The difficulties posed by this lack of domestic supply were made particularly acute during the First World War when oil became, for the first time, vitally important as fuel for the cars, trucks, tanks, submarines and aeroplanes that had begun to define modern warfare. German problems securing an oil supply were a serious hindrance to their war efforts, and the importance of oil to the eventual outcome of the war was highlighted by Lord Curzon in a speech made 10 days after the Armistice, in which he stated that 'The Allied cause had floated to victory upon a wave of oil.'[20]

During the First World War, and throughout the 1920s, there were concerted efforts by both European countries and the USA

to gain control of world oil supplies. The former Ottoman Empire was a particular focus, with British and French attempts to control deposits in Iran and Iraq directing many of their wartime actions in the region.[21] After the formation of the Iraq Petroleum Company in 1928, which essentially saw concessions in the area licensed by a cartel of British, French, Dutch and US companies, the search for Middle Eastern oil extended to Saudi Arabia. At the same time international oilmen were flocking to Latin America to seek concessions in the booming markets of Mexico and Venezuela.[22]

Oil had become a significant player in global power, with increasingly elaborate systems developed to control its movement across national borders and through global trading networks [90]. Pipelines and oil tankers – both of them transport systems that had been developed in the nineteenth century – grew in scale and geographic spread. In the popular imagination, oil companies linked their products with national progression through extensive advertising campaigns that celebrated petrol as a means of exploring the modern nation by car [91].[23]

The USSR, which had a large domestic supply through the oil fields at Baku, demonstrated the vast political significance that petroleum had taken on by the mid-1930s in a series of advertisements drawn for the state tourism agency, Intourist. Baku was here promoted as a destination for Soviet citizens, its oil industry celebrated in a poster foregrounded with an image of a giant derrick [92]. No longer mediated by the experience of driving patriotically through nation or empire, the awed and celebratory relationship between oil and national power was here laid out in the clearest possible terms. As the geographer Alexander Radó summed it up in 1938, in his *Atlas of To-day and To-morrow* [93]:

The expansion of motor and air transport in the early part of the 20th century marked the beginning of a triumphal march for petroleum, most important of the liquid fuels. It represents

↑ **89** Panoramic photograph, by F.J. Schlueter, c.1919, showing the Goose Creek oil field, one of the richest fields in the 1910s Texan oil boom

→ **90** Laying the Iraq Petroleum Company's pipeline across the Plain of Esdraelon, by the Photographic Department of the American Colony in Jerusalem, July 1933. The pipeline connected oil fields at Kirkuk, in Iraq, with refineries and ports at Haifa and Tripoli

STONEHENGE
SEE BRITAIN FIRST ON SHELL

20 percent of the world's total mining production and is thus the chief mining product after coal; it also provides about a quarter of the power in the world's economy. It is of great military importance for the air forces, navies, the motor transport and tanks of modern armies, and as it is almost entirely lacking in the territories of most Great Powers, it is the most fought for commodity in world trade.[24]

'Petroleum helps to build a better life'[25]

While the foundations of an oil economy were clearly laid out in the 1920s and '30s, the market for petroleum reached its peak after the Second World War. Driven directly by a massive increase in car ownership, demand for gasoline shot up.[26] Production was able to keep step with rising consumption due to increasingly sophisticated techniques for drilling, transportation, refining and prospecting that had been developing steadily since the 1920s.[27] By 1964 oil had surpassed coal as the world's most commonly used fuel, and by 1970 rich countries were consuming four times as much oil as they had in 1950.[28]

↑ **91** Stonehenge poster, by Edward McKnight Kauffer, 1931, from the 'See Britain First on Shell' advertising campaign, selling modern, national tourism by car

→ **92** Poster promoting tourist travel to the Soviet oil fields at Baku, produced for Intourist, c.1936

BAKU

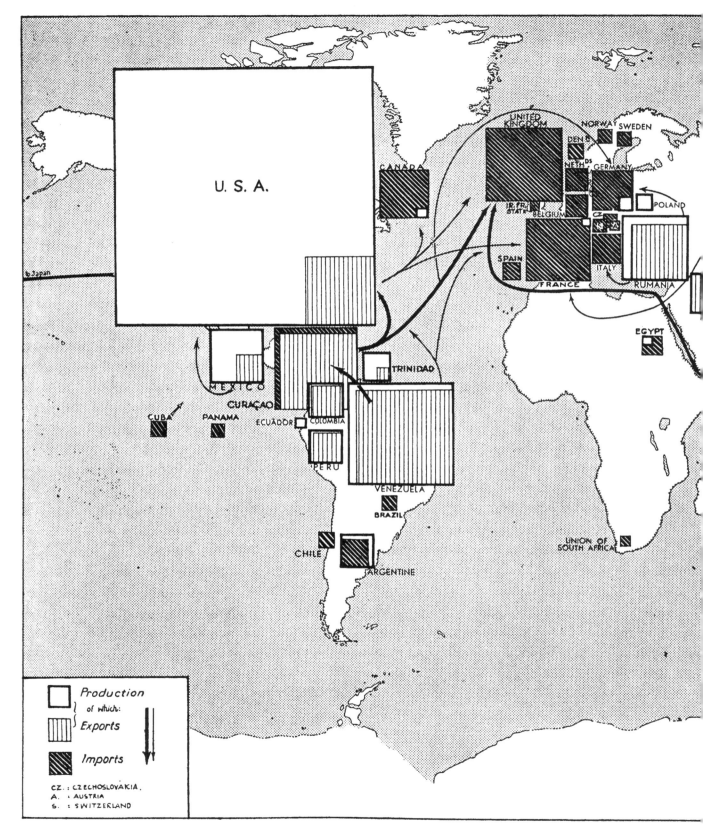

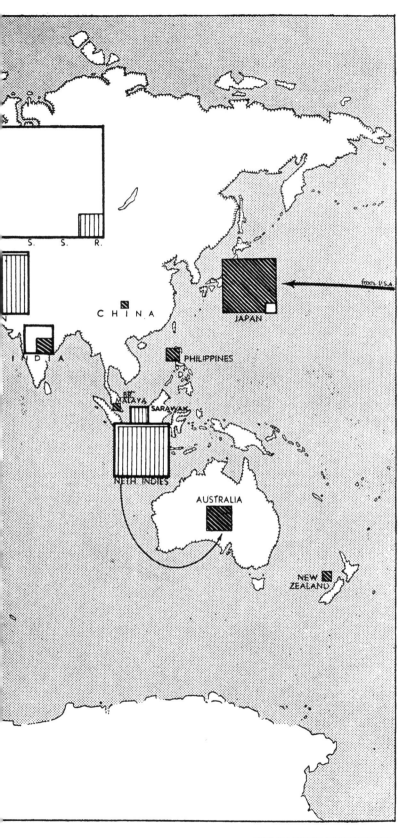

← **93** Map showing the
quantities of petroleum
that were produced and
consumed by different
countries around the world,
published in Alexander Radó's
*The Atlas of To-day and
To-morrow*, London 1938

Major oil companies, floating in fuel for most of the 1950s and '60s, were keen to sell petroleum as the key to a new way of living. This began with the car, with prices for crude oil being dropped in 1960, and cheap fuel enabling US manufacturers to produce ever larger and more petrol-hungry models. Giant, gas-guzzling engines became the norm, heralding a huge shift in fuel efficiency: while a large Cadillac would have driven around 20 miles to the gallon in 1949, by 1973 American-made cars averaged only 13.5 miles per gallon.[29] At the same time that gasoline consumption was increasing, oil companies also competed to create high-octane fuel that would give drivers more horsepower. Selling the vision of a new kind of freedom on the open road, the oil industry encouraged and enabled the massive increase in speeds that were built into the design of post-war automobiles (see pp. 50–2).[30]

Beyond its engine, the petrol car also became the point of origin for a much wider world of petroleum products. On the back of advances in refining techniques made during the war, manufacturers were now able to produce a vastly increased number of plastics and petro-chemicals. The post-war oil industry sought to capitalize on this by creating a vision of a swarm of new petroleum products, saturating and improving every aspect of life.[31] Starting with the car, this wave of bounty would stretch out to encompass everything from the plastics in a new suburban house, to furnishings, clothing, cleaning products and garden fertilizers. As Standard Oil framed it in 1957: 'Most of the rubber we use in this country is made from oil, manmade fibers are derived from oil, asphalt roads, medicines, all made from that incredible wonderbox petroleum.'[32]

This image of plenty had clear implications for the Cold War climate into which it was born. It was a vision that the USSR sought to match, with a Soviet poster designed in 1960 making the link between petrol, power and prosperity crystal clear [94]. An image of oil gushing from a pipeline is here strikingly mirrored by a flow of coloured plastics, while the text under the graphic reads: 'From oil we take for the needs of our country a river of gasoline, oil and petroleum and in addition thousands of items for the home and for domestic comfort!'

Creating an impression of an endless supply of fuel – a national bounty that will never run out – the poster's graphic was reflected materially in the extravagant cars, and the cascade of petroleum-based products, that were manufactured around the capitalist world during the 1950s and '60s. A sense of boundless optimism encompassed these decades, filled with awe at the technological power that humans had harnessed in order to mould their natural environment. Nowhere can this be seen more starkly than in a 1962 advertisement for the Humble Oil Company [95]. Proudly proclaiming that 'To meet the nation's growing needs for energy, Humble has applied science to nature's resources', the advertisement spins around a photo of a glacier and a staggering claim: 'This giant glacier has remained unmelted for centuries. Yet, the petroleum energy Humble supplies … could melt it at a rate of 80 tons each second!'

→ **94** 'From oil we take for the needs of our country a river of gasoline, oil and petroleum and in addition thousands of items for the home and for domestic comfort!', poster designed by Viktor Koretsky, c.1960

→→ **95** Advertisement for the Humble Oil Company, published in *Life* magazine, 2 February 1962

EACH DAY HUMBLE SUPPLIES ENOUGH ENERG

This giant glacier has remained unmelted for centuries. Yet, the petroleum energy Humble supplie
converted into heat—could melt it at the rate of 80 tons each second! To meet the nation's growing n
for energy, Humble has applied science to nature's resources to become America's Leading Energy Comp
Working wonders with oil through research, Humble provides energy in many forms—to help heat our ho
power our transportation, and to furnish industry with a great variety of versatile chemicals. Stop at a Hu
station for famous Esso <u>Extra</u> gasoline, and see why the "Happy Motoring" Sign is the World's First Ch

'Each day Humble [Oil] supplies enough energy to melt 7 million tons of glacier!'[33]

As the Humble advert so painfully shows, this was a moment of incredible – and catastrophically destructive – hubris. The petroleum-filled Aladdin's cave began to close its doors in the 1970s, initially through a wave of oil-company nationalizations triggered by the formation of OPEC (the Organization of the Petroleum Exporting Countries) in Baghdad in 1970. Under OPEC's founding agreement, oil-producing member states agreed not to tolerate further lowering of crude-oil prices by major multinationals. World oil prices quadrupled between 1973 and 1975, sending shock waves through North American and European markets. After decades of price-setting and remote control, Europe and North America faced nations that were unsympathetic to their demands – following the October 1973 war with Israel, Arab oil-producing countries announced production cuts to be put in place until Israel withdrew from occupied Palestinian territories. The resulting 1973–4 oil crisis saw restrictions on petrol at the pumps throughout Europe and the US, fuelled by panic about a possible impending shortage [96].

At the same time that the global geography of oil was shifting, there was also increasing alarm about the vast environmental destruction wrought by the gasoline engine. This ranged from disastrous spills out of a growing network of oil rigs, tankers and pipelines, to the increasingly urgent problem of carbon emissions and air pollution. Although manufacturers had been aware of the huge rate of automobile emissions since at least the 1950s (and even 1890s accounts of petrol motors often discussed their 'dirtiness'), there was no action

Think of it as a supermarket that gets 42 miles per gallon.

The 1975 Honda Civic CVCC. You can take it to the gas station or the supermarket and not be disappointed.

In EPA lab tests for highway driving, the 5-speed Civic got 42 miles per gallon. The best mileage performance for any car sold in the U.S.

Yet, we didn't sacrifice roominess for economy. The Civic seats four adults. It also holds a lot of groceries, or luggage, or sporting gear, or you-name-it. (And that's with the rear seat up. Fold it down and there's even more carrying space.)

The secret behind this kind of performance is in the way the Civic was designed. We gave it a shorter profile so its compact size would be ideal for today's driving. We gave it a lot of interior room for today's life-styles.

And finally, we gave it an engine that doesn't need a catalytic converter, that runs on regular, low-lead or no-lead gas: the Honda CVCC Advanced Stratified Charge engine.

All in all, it's one car that no matter where you go, you won't mind saying those three little words: 'fill 'er up.'

CVCC and Civic are Honda trademarks. ©1975 American Honda Motor Co., Inc.

HONDA CIVIC
What the world is coming to.

96 Petrol station attendants in Portland, Oregon, during the 1973 oil crisis, photograph by David Falconer. This is part of 'Project Documerica', a 1970s scheme under which the newly formed Environmental Protection Agency hired photographers to capture images of environmental problems and everyday life

↑ 97 Advertisement for the Honda Civic, emphasizing the car's fuel efficiency, 1975

to limit this until a wider public engagement with the issue in the late 1960s and '70s. With 60–80 per cent of US air pollution deriving from motor vehicles in the mid-1960s, a number of state and eventually federal measures were introduced to force car manufacturers to reduce emissions. Even after the Clean Air Act was passed in 1970, it took most of the rest of the decade for many US manufacturers to make any real changes to the levels of pollution from their cars.[34]

Despite this reluctance to adapt engines for lower emissions, shifts in the climate and control of oil did eventually force a change in automobile design. In response to a growing awareness of air pollution, Ford, General Motors and other companies such as General Electric made highly publicized attempts to design battery-powered cars in the mid- to late 1960s. General Motors produced the 'Electrovair' – an adaptation of the Chevy Corvair – which was powered by a zinc battery. Although the car's small range, and the heaviness of the battery, meant that it could never become successful in production, the design did give General Motors some traction when asked to offer proof of a company response to air pollution.

More significant than these short-lived electric experiments was a consumer-led shift away from large cars. Where giant gas-guzzlers had previously reigned supreme on the US market, drivers hit by increases in fuel prices began to buy smaller cars in the early 1970s. This triggered a growth in the sales of fuel-efficient Japanese and European models, prompting American manufacturers to scale down, in an attempt to match the competition [97].

Unfortunately these changes were very temporary. With the collapse of high oil prices in the mid-1980s, and an economic boom in the 1990s, drivers were quick to return to larger, petrol-hungry models. Huge gas-consumers like SUVs gained a massive proportion of the global market and, according to the UN, energy use for transport (of which 95 per cent was petroleum) increased by 66 per cent between 1973 and 1996.[35]

As we move into the third decade of the twenty-first century we inhabit a world in crisis. In October 2018 the UN released a climate report warning of the now-unavoidable and disastrous consequences of global warming. Without extremely drastic measures in the next 10 years, the planet is facing huge rises in sea levels, a devastating loss of biodiversity through species extinction, spreading areas of drought and large-scale loss of forest.[36] While the combustion engine is not the only factor that has fuelled this fire, petrol cars remain an extremely significant contributor to carbon gas emissions.[37]

Triggered by this dire environmental situation, electric vehicles are beginning to increase in number and profile. Their new fashionability was perhaps best illustrated by the wedding of Meghan Markle and Prince Harry in spring 2018, when millions of viewers watched as the happy couple drove away from their reception in a Jaguar E-Type fitted with an electric engine [98]. This was a master stroke in public relations and image-making, weaving together entrenched British icons (the monarchy and the E-Type) with symbols of a progressive forward-looking nation (a mixed-race couple and electric energy).

↙ **98** Meghan Markle and Prince Harry leaving their wedding reception in an electric Jaguar E-Type, photograph by James Gourley, 2018

→ **99** 'Tiny House Warriors' – Secwepemc nation protestors opposing the Kinder Morgan pipeline, 2018

Despite the seeming optimism of the gesture, the E-Type's neat splice between forward-thinking and nostalgia does point to murky waters ahead. Jaguar's decision to encase their electric future in the body of a speeding 1960s sports car highlights the difficulties inherent in moving on from the dream of gasoline. Although electric cars are beginning to be produced in larger quantities, with manufacturers paying greater attention to the need for an alternate fuel, petrol cars remain hugely dominant and the non-petrol engine is still (as it was in the 1890s) very often discounted due to concerns about battery range and the lack of a refuelling infrastructure.[38]

It would appear that the same culture that shaped nineteenth-century attitudes to electric and petrol cars continues to have an influence. The long life of the so-called Jeep 'Cherokee' (now approaching its 45th year of production) highlights the extent to which manufacturers persist in selling drivers the myth of the wilderness, and a vision of heading out to conquer the unknown. First Nation protests, such as that currently being led by the Secwepemc nation against Canada's Kinder Morgan pipeline, have recently highlighted and disrupted the often invisible networks of power, wealth and dispossession that have accompanied so much of the history of the gasoline car [99]. Despite this, Citroën's idea of striking out in a speeding automobile, limited only by a supply of petrol, remains a tantalizing inheritance, and one that continues to be claimed as a right by much of the world's population.

WATER WARS AND THE MIRACLE METAL

Laurence Blair

An ancient, dried-up seabed nearly four kilometres above sea level stands in the north-west corner of Argentina, on the Andean *puna* (plateau) of Jujuy Province, which stretches between the Chilean and Bolivian border. Standing on the blinding-white shores of the vast Salar (salt flat) de Olaroz, it isn't obvious that a billion tonnes of low-grade lithium ore are suspended in brine for one kilometre beneath the salt – or that a newly installed, multimillion-dollar industry is extracting it, evaporating the water and refining the metal. But defy the thin air and climb one of the parched brown hills that ring the valley – or pull up and consult satellite images – and the futuristic warehouses, and 18 long, curved evaporation ponds ranging from emerald to aquamarine, are impossible to miss. A shared venture between Canadian–Australian firm Orocobre, a subsidiary of Toyota Tsusho, and the provincial government, the Salar de Olaroz provides 10 per cent of the world's lithium demand – and is set to double production in the next two years.[1]

Olaroz is one of a handful of vast salt flats – scattered between the high-altitude deserts of north-west Argentina, south-west Bolivia and northern Chile – that together make up the lithium triangle. This arid, inhospitable corner of South America was for centuries the sole preserve of semi-nomadic herders and passing trains: both caravans consisting of llamas and mules and, latterly, locomotives carrying copper and other metals. But it has become the focus of a new, high-tech gold rush in little more than a decade. The triangle is believed to contain more than half of the world's stocks of lithium – the ultra-light mineral used to make lithium-ion batteries. These are an essential component in multiple technologies of the future, including solar batteries, smartphones, laptops and, above all, electric vehicles (EVs). With EVs expected to make up over half of all new car sales, and 33 per cent of global vehicles, by 2040 – adding up to 530 million EVs worldwide – demand and prices for lithium are soaring.[2] Multinational mining companies are in turn flocking to the lithium triangle, bringing employment and economic growth, and often investing in local hospitals,

schools and community tourism initiatives as part of their concessions. The industry is sometimes presented as wholly benign, with EVs representing a break with the fossil-fuel economy, its ruinous effects on the climate and historical cycles of commodity boom and bust. In line with these high expectations, lithium is sometimes dubbed 'the miracle mineral'. The end product – the sleek design of a Tesla Model X or a BMW i3 – may seem sustainable, disassociated from any side effects, but the extraction of lithium, and other metals like copper and cobalt, for EV batteries still obeys the traditional logics of geography, economics, politics and ecology. Lithium mining is already having a significant impact on local environments, the severity of which is yet to be completely understood – or fully divulged. The question of who benefits and who loses out is also fuelling societal divisions and change, and unequal conflicts between local activists and international forces.

The small town of Susques – around 90 minutes' drive from Olaroz – is host to perhaps the starkest of these contests. Little more than a few dusty streets of brick bungalows, a queue of slumbering long-distance trucks bound for the Pacific, a military radar base and a sixteenth-century straw-and-adobe church with a cactus-wood ceiling, the town sees temperatures falling below freezing every night. Much of the population spends days at a time in small huts out on the *puna*, rearing llamas and goats. The mines have brought new prosperity to some, with hefty 4WD trucks parked outside the houses of company employees. But Carlos Guzman, an indigenous Atacameño pastoralist – who moonlights as an overall-clad technician at the municipal generator, where I met him – is among the critics. In 2014 he co-founded Colectivo Apacheta, a group of a dozen families who came together to challenge the newly active lithium mine. Their main fear, he said, was that the heavy water usage of the plant – up to 200 litres of brine is evaporated per second, his figures suggest – would kill off local vegetation, and hence their livestock and eventually their livelihoods. A public consultation held by Minas de Olaroz in Susques secured only a few dozen signatures out of a population of 2,000 – some of them faked, Carlos alleged. After several of the biologists and engineers brought to survey the possible environmental impact of the mines disagreed with the decision to give them the green light, their findings were not released. Attempts at dialogue with the multinational fizzled out. 'We don't understand fully how the process of purification of lithium carbonate works,' Carlos admitted. But the collective intuited that there had to be some impact, in part inspired by local beliefs that water is the blood of Pachamama, mother earth. For Carlos the results are already plain to see, in the tough, spiky grasses and plants that feed his llamas and goats. 'Last year there was no rain. In 2000, 2001, the drought was a lot worse. But our plants managed. In this single year of drought [2016], the plants ...' He mimed a wilting, a collapse. As a result, his animals starved, their ribs showing through taut, stretched hides. Local plants have up to 10 metres of roots, to draw water from beneath the arid earth, he explained, so underground aquifers may already be severely depleted.

The impact goes beyond the immediate area, he argued. 'Everything they don't need – mud, sand, chalk, caustic soda, they put back into the *salar* [salt flat]. They're contaminating above and below.' Polluted water systems could poison urban populations in the nearby cities of Jujuy and Salta, he suggested. The mass evaporation of water could already be contributing to heavy rainfall and flooding in the agricultural provinces of Argentina's pampas, far to the south-east.[3] 'The disequilibrium is huge,' he emphasized. Carlos had little hope that Colectivo Apacheta would prove victorious over market forces and political interests. 'Of course, the result will be negative for us, because we're a minority with little money. It's logical,' he said. Yet he invoked the multiple meanings of *apacheta*, a noun common to both Quechua and Aymara. It describes the small cairns often seen on high Andean passes, formed stone-by-stone by passing travellers since at least the time of the Inca: an offering, a signal, a waymarker. 'We want to leave this written and set down for the world,' he concluded. 'If we don't look after our nature, we're all dead.' A few sandy streets away, Elva – Carlos's sister,

another founding member of the group – similarly sketched their clash with the lithium companies as a fight between tradition and the future. 'As an indigenous people we've looked after the water for thousands of years,' she said. 'We lived in the desert, we respected each other and nature, not like now. The world itself has technologized ... there's no respect for Pachamama.' Minas de Olaroz has insisted that its activities leave a minimal environmental footprint, pointed to its investments in the local community and denied any wrongdoing in the process of establishing local operations.

In the quaint colonial city of Salta, 300 kilometres to the south-east, I visited a lithium mining firm active in the salt flats of the surrounding province. Its headquarters were in a converted one-storey house on a residential street. Seated at the kitchen table, an Australian executive complained that activists and journalists had demonized external firms and exaggerated, without firm evidence, the environmental impact of mining. He emphasized that his company was working closely with local communities. Ultimately, he insisted, the company complied with the province's environmental law to the letter: it was up to politicians to toughen up restrictions and monitoring, if they thought it in the public interest. Carlos had agreed: rather than with mining firms, which – after all – were only fulfilling their function, the real responsibility rested with shortsighted politicians, who were prepared to overlook permanent environmental and social changes in exchange for a few decades of economic gain.

Argentina's plans for lithium extraction on the *salares* are a key part of a broader development plan for this long-neglected and predominantly poor region. In September 2018 the provincial government of Jujuy broke ground on a new solar-energy plant – the largest in the world – on the *puna* near Olaroz and Susques. The 'Plan Belgrano' of centre-right President Mauricio Macri has promised US$16bn of investment in airports, housing, highways and railways across Argentina's north-west.[4] When I passed through the area in late 2017, workers in orange overalls were tearing up British-built track laid down in the 1940s. The plan is to

↑ **100** A vehicle turns on to the road adjacent to the Salar de Olaroz. Minas de Olaroz has invested in local services. But local activists fear the environmental consequences of lithium mining

reconnect the railway from the city of Jujuy, via the rocky Quebrada de Humahuaca, to the border with Bolivia, some 300 kilometres to the north. But the rusting leftovers of British railway heritage dotted throughout the valley – part of an industry that once shipped sugarcane from vast haciendas to the ports of the Río de la Plata – demonstrate how previous commodity booms have come and gone without leaving lasting wealth behind. Carlos had little hope that the lithium boom and its trickle-down effects would linger. 'What they've invested in comparison with what they're taking out – how much are we talking, percentage-wise? Not even 0.001 per cent. Our regional economy, based on llamas, multiplies over time depending on their care,' he added. 'Mining is only an investment in contamination. The 250 workers are there today, tomorrow. But once the lithium has gone, where will they go?' Such caution is likely to fall on deaf ears. With lithium one of the few bright spots in the country's crisis-prone economy, Argentina is set to double down on extracting the white gold in years to come – for better or for worse. Macri has scrapped an export tax on mineral products

and cancelled a ban on companies sending profits overseas.

Further north in Bolivia, too, plans to extract lithium are entwined with politics, economics, history and ecology – even as Bolivia's leaders are ideologically opposite to the current administration in Buenos Aires. The Movement towards Socialism (MAS) government of the indigenous Aymara president, Evo Morales, has high hopes of tapping the vast reserves of the metal under the 10,000 square kilometres of the Uyuni salt flat. Bolivia's lithium stocks are 'the largest in the world' and the extraction process is 'very advanced', Morales promised a crowd of *campesinos* in 2018: 'Soon Bolivia will set the price of lithium batteries around the world.'[5] But most observers are more cautious: analysts put Bolivia's share of the world's lithium at 25 per cent, and note that in 2017 the country produced just 120 tonnes of lithium carbonate per year, far less than Chile (70,000 tonnes) and Argentina (30,000).[6] Foreign expertise and investment have been scared off by Morales's history of fiery rhetoric and partial nationalization of the oil and gas industry soon after his election in 2005. A deal was signed in

↑ **101** A child crosses abandoned rail sidings at the village of Colchani, on the fringes of Bolivia's Salar de Uyuni. Bolivia's government hopes lithium mining will bring investment to this poor region

April 2018 with ACI Systems GmbH, a German company, to invest US$1.3bn in Bolivia's fledgling lithium industry, from extraction through to the creation of batteries and cathodes. If production can pick up substantially by late 2019, it would add weight to Morales's claims to have delivered an economic and social revolution to Bolivia. This might be enough, perhaps, to let him scrape a fourth consecutive electoral victory in 2019, despite sustained allegations of corruption and authoritarianism. But few believe that Germany is 'at the leading edge of technology', as Bolivian analyst Juan Carlos Zuleta told *Americas Quarterly* magazine in May 2018. 'By now most people know where the tech is. It is in the US with Tesla.'[7]

There are also fears that the industry's heavy water usage could have unforeseen consequences for a corner of Bolivia already beset by drought. In the adobe-brick village of Colchani on the edge of the Salar de Uyuni, I spoke with Aureliano Mauricio Valero, formerly a fisherman from an indigenous Urus-Muratos community on the shores of Lake Poopó, 150 kilometres to the north. When the lake disappeared in late 2015 – linked, say experts, to

global climate change, mining pollution and overuse of the water – his family migrated here for good. Inside a dimly lit shed, Mauricio, his wife and daughter scooped salt into plastic bags. Together, they could pack 5,000 bags a day, earning 125 bolivianos (£14). Outside on the dazzling salt flat, a few dozen neighbours who had emigrated with them hacked grey-white bricks out of the ground by hand. Aureliano recalled fishing as a boy, making Lenten offerings of sweets to the water and casting his nets through the night. 'We enjoyed working,' he said. 'Our work is Lake Poopó, and with that dried up we're like orphans.'

The disappearance of what was once the country's second-largest lake – and the mass die-off of some 200 species of birds, mammals and fish, rotting where they lay – is testament to the fragility of the climate in the south-west of Bolivia. It is not difficult to imagine the extraction of lithium brine, depleting vast underground aquifers in the process, similarly having far-reaching impacts well beyond the Uyuni area. Local conservationist Manuel Olivera has warned that the government has yet to consult local communities about the

↑ **102** Spiky, tough grasses – eaten by the llama herds owned by locals – dot the roadside near the Salar de Olaroz, drawing water from underground reservoirs to survive in the desert environment

environmental impact of lithium mining. When the former coca-leaf grower and union activist was sworn in as president in 2006, Morales promised 'an end to injustice, to inequality ... from 500 years of Indian resistance we pass to another 500 years in power'. European colonization and neoliberal economics had led to nothing but the 'looting of our natural resources': from the silver of Potosí, the churning motor of Spain's imperial economy, to the rubber of the Amazon, poured out and stretched to make tyres for American Jeeps in the Second World War.[8] Yet while the profits of Bolivia's nascent lithium industry may be reinvested to help the country's poorest, at a local level the impact of the trade may prove to be as ephemeral as the plundering that preceded it.

Across the Andes is the third and most significant corner of the lithium triangle: Chile, home to as much as 50 per cent of the world's reserves of lithium, much of it suspended in brine beneath the Salar de Atacama. Here, the struggle between lithium and copper mines on the one hand, and local communities and environmentalists on the other, is nothing less than a 'water war', in the words of a local NGO, the Atacama Desert Foundation.[9] Major industry players Albemarle and SQM, the world's largest lithium producer, have dominated local production for decades. But the long-established presence of lithium mining means that locals and NGOs have had time to organize, as well. They complain that diverting freshwater aquifers and streams to top up the brine of the *salar* has led to shortages of water for drinking and livestock, and is probably causing unknown knock-on effects across the region. SQM was assigned groves of Algarrobo trees – tough desert hardwoods with deep underground roots – near its installations to monitor, serving as a 'canary in the mine' for water shortages. Starting in 2013, government inspectors found that one-third were shedding their leaves and dying.[10] In mid-2018 Chile's water regulator announced plans to ban mining firms from extracting water from several key aquifers. But political, economic and geographical factors are set to make Chile the world's largest lithium producer once again, having briefly ceded that position to Australia in 2017.[11]

The centre-right government of Sebastián Piñera, which returned to power in January

2018, sees lithium as a strategic resource that is key to Chile's development, particularly as global demand and prices for copper (the country's principal export) stall and the grade of copper ore in Chile's north declines. While Australia's lithium is embedded in rock, the brine suspension of Chile's *salares* makes lithium extraction comparatively easy. And the Atacama's proximity to the Pacific, and shipping lanes to key markets like China – as well as the lower altitude relative to Argentina and Bolivia – put market-friendly Chile in pole position. SQM is due to increase its lithium production to 70,000 tonnes in 2018, rising to around 150,000 tonnes of lithium carbonate equivalent (LCE) by 2021, nearly one-third of the global market.[12] There are also suggestions that the manufacture of EV components could leap out of the confines of the lithium triangle, drawing other South American countries into a pan-regional green supply chain, transforming economies and societies along its route. A White Paper published by Duke University in June 2018 suggested that lithium, copper, manganese and cobalt from Chile, Argentina and Bolivia could be shipped to Paraguay, where the abundant renewable energy from the Itaipú Dam – the world's most powerful – could be used to manufacture batteries. These could then be exported to the automotive manufacturing hubs of Brazil to produce EVs.[13] Yet matching and challenging this international dimension are new links between the ancient communities that inhabited the lithium triangle long before post-colonial borders divided it in three. Elva Guzman told me that Colectivo Apacheta had met their fellow indigenous activists in Chile and Bolivia to share information and strategies. 'What are we going to do without water? Where will we go? This is what scares us, and why we're fighting,' she said. The simple plan was 'to defend what is ours. To not give it away. Because I can't go anywhere else. These are my roots.'

There are voices of dissent and warning signs amid the wild optimism. One research firm warned in early 2018 that the lithium industry was 'sleepwalking into a tsunami of oversupply', as exploding lithium production starts to outstrip demand. Prices in China, the world's biggest lithium consumer, nearly halved between March and August 2018, falling from a historical peak of US$24,750 per tonne to US$13,000.[14] Investment banks, and even SQM itself, predict demand barely matching half of supply by 2021–2. The industry will probably self-correct and adjust to such bumps in the road. Yet the Atacama is littered with the eerie detritus of previous commodity booms and collapsed industries: a warning, perhaps, of what might come. The highway from the Salar de Atacama to the Pacific is dotted with abandoned nitrate-mining towns from the early twentieth century: Caracoles, Chacabuco, Pampa Union, Baquedano – rusting furnaces, clapboard houses invaded by sand, looted cemeteries littered with scraps of bone and fragments of clothing, graffiti and murals left by the artists and activists incarcerated here under the Pinochet regime. The silent rooms and mouldering rail yards invite reflection on what the remnants of today's industries – the economic lifeblood for thousands, and part of a supply chain serving millions – will look like if (or when) they are eventually abandoned. Even industry insiders give most current lithium-triangle projects a shelf life of just 40 to 50 years before the mineral is depleted.

Less labour-intensive than the commodity booms that preceded it, the lithium trade will leave no sprawling company towns – perhaps just a few abandoned Portakabins or squat desert bunkers. The reservoirs will dry up and crystallize, no longer the iridescent jewels amid the sand once seen from above. The shattered glass and warped chrome of visitor centres will form a new geography of ruin alongside the old one of adobe-and-straw huts. The futuristic refineries will linger in the backdrop to action films, in the abstract artworks funded by multinationals and in promotional videos on YouTube featuring tanned, helmeted workers in UV sunglasses. Perhaps the years of growth and investment that the mines bring will stick, leaving prosperous communities in their wake. Or perhaps the shores of the *salares* will fall quiet once more: only now without water, plants, animals, people. A perfect desert.

103 & 104 Aureliano Mauricio Valero, his wife and daughter earn £14 a day packaging salt from the Salar de Uyuni. Formerly a fisherman, Valero's livelihood dried up due to climate change

105 Indigenous Atacameño pastoralist Elva Guzman is a founding member of Colectivo Apacheta, which is resisting lithium mining on the Salar de Olaroz

→ **106** Evaporation pools at the Rockwood Lithium Mine in Chile extract lithium salts from brine. Global demand for lithium is soaring, fuelled in large part by the growth of the electric vehicles market. Still taken from *The Breastmilk of the Volcano* (2018), directed by Unknown Fields

DRIVING

Brendan Cormier

THE

NATION

↑ **107** 'Halftime in America' is a Chrysler TV advertisement that aired during the 2012 Super Bowl. It starred Clint Eastwood delivering a speech comparing America's car industry with the strength and character of America itself

On 5 February 2012 a peculiar car commercial aired during America's most-watched sporting event, the Super Bowl. The coveted slot had generally been given over to adverts competing for humour, irreverence and big-budget flare. This commercial, however – called 'Halftime in America' [107] – stood in complete contrast; it was sombre and deadly serious. It opens with Clint Eastwood, the icon of American cinematic grit, obscured in shadows while walking through a dark tunnel. In a speech laden with dramatic sports metaphors, he recounts how the country had been brought to its knees by the Great Recession; that good American people were suffering; and that a 'fog of division, discord, and blame made it hard to see what lies ahead'. The commercial is overlaid with images of classic Americana cast in desaturated tones: single-family homes and industrial landscapes, a bird's-eye flyover of Manhattan, followed by the image of morning dew over a lonely rural porch. We see portraits of everyday Americans doing everyday things: getting out of bed, dropping the kids off at school, working in factories, rowing on the water, hanging out at the pier. Sweeping minor chords from distant horns and strings help to build the drama, underscoring Eastwood's finale as motivational speaker, suggesting how the American auto industry is about to lead the way into its 'second half'. His closer is both rousing and menacing, coarsely delivered in his characteristic growl: 'The world's about to hear the roar of our engines.'

The commercial became a viral hit, was lampooned on the weekly comedy sketch show *Saturday Night Live* and was roundly criticized by Republicans for being overtly political and congratulatory of the bailouts that the industry received following the financial crisis.[1] Although breaking from Super Bowl form, however, the messaging reinforced a long-held and deeply entrenched understanding shared by many: that American industrial might was a huge part of the country's identity, and that the auto industry was indelibly at the centre. This gets to the core of a curious phenomenon in the history of the car: it had the fate of coming into existence at the same time as another hugely important late nineteenth-century innovation, the nation state. And, as we shall see, the two did not simply coexist; rather, they developed a highly symbiotic relationship with each other: nations helped build cars, and cars helped build nations.

Connecting a Country

For cars to become successful, they first needed places to go. While rail networks helped modernize empires in the nineteenth century, it was through national highway construction programmes that many burgeoning twentieth-century nation states sought to modernize. At the beginning of the twentieth century the number of limited-access highways in the world was zero; the first – the Long Island Motor Parkway – was built in 1908.[2] Today, in contrast, hundreds of thousands of expressways criss-cross the world, making highways collectively one of the largest building projects in human history. These networks have fundamentally altered the way people and goods move around, the way economies are governed, the way disparate geographies are connected, and the shape of a nation.

The man who helped to popularize the idea, however, was not a calculating politician, but a savvy Italian entrepreneur, Piero Puricelli. He was an avid motorist, with strong connections to the local Milan business elite, and a specialist in road construction. Following the First World War, guided by his own professional self-interest, he became a major advocate for road renewal and the concept of the limited-access highway. In 1921 he prepared a pamphlet that proposed a motorway connecting Milan to Lake Como, Varese and Maggiore – a kind of road that had never before been seen in Europe, on which only automobiles were allowed access.[3] He sold his futuristic vision on the premise of savings and speed: language that spoke well to the local business community. After creating substantial hype for the proposal, in 1922 it caught the eye of the newly appointed prime minister, Benito Mussolini. He decided to back it, by dedicating

→ 108 The Autostrada leading from Milan to Lake Como, shortly after its opening in 1924

state funds to the private venture and passing laws that enabled the necessary expropriation of land, and by 23 March 1923 Mussolini could be seen participating in the ceremonial first strike of the pick, symbolizing the commencement of construction [108].[4]

For the fascist government, Puricelli's vision of futuristic high-speed roads connecting the country served not only savings and speed, but also as perfect propaganda for the new regime. The 84-kilometre Milan–Como highway opened in 1924 and quickly sparked a European-wide enthusiasm for new highway plans and projects – helped along by Puricelli's panache, as he criss-crossed the Continent selling the story of his domestic success. Suddenly Italy was thrust into the spotlight as a model of modernity to be followed. During this period ambitious new highway plans were drawn up in France, the Netherlands, Spain and the UK, while back home, pro-posals for other highway projects came streaming into Mussolini's office.[5] As the country had unified as a nation state only in 1861, it still struggled to connect its diverse regions and cities. Puricelli's motorway concept was seen by many as the perfect antidote to a divided nation [109]. But, curiously, despite its powerful propaganda function, the fascist government soon grew wary of the financial risk such capital projects entailed, and only reluctantly approved a handful of other motorway projects from then on.

By 1935, with the completion of the Genoa–Serravalle truckway, Italy had built almost 500 kilometres of motorway[6] – a feat that might still impress, had it not been vastly eclipsed in just six short

← **109** A map published by Piero Puricelli in 1933, showing the condition of roads across Italy. The segments illustrated as double parallel red lines indicate the completed Autostrade of the time

→ **110** The newly opened Autobahn from Berlin to Szczecin, early 1930s

years by its northern ally, Germany. The Third Reich took the early lessons from the Italians, as seen in Como, and scaled-up highway production to massive proportions. In 1933 it unveiled its scheme to build 6,000 kilometres of four-lane *Autobahnen* in the span of just five years [110]. An entire nationalist rhetoric was built around the project through pamphlets, newsreels and bold claims, dubbing the roads 'the pyramids of the Reich'. The narrative included ideas about linking the German people (*Volk*), and connecting urban areas (which were much maligned by the Nazi Party) with the countryside (considered wholesome and restorative). It was also touted as a massive make-work programme, promising to employ 300,000 out-of-work Germans, giving a substantial boost of cash – five billion Reichsmarks – to the national economy.[7] Those claims, as it turns out, were widely exaggerated, but Germany had nonetheless set a new precedent for coordinated highway planning at a national scale, which would be taken up by others following the Second World War.

In the 1950s, with booming economies and a mobilized technocracy, highway-building in Europe finally took off on a scale that early

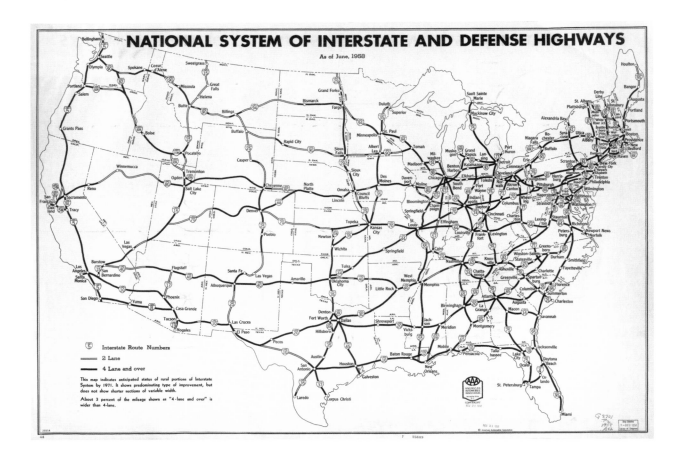

NATIONAL SYSTEM OF INTERSTATE AND DEFENSE HIGHWAYS

As of June, 1958

advocates like Puricelli had only dreamed of. In Italy 1,300 kilometres were built by 1961, 4,300 by 1971 and 5,900 by 1980.[8] In Britain the Special Roads Act 1949 allowed for the first time the construction of limited-access highways, leading to a flurry of road-building, starting in the late 1950s, which would continue for decades.[9] In 1956 the USA unveiled its own ambitious Federal Highway Act, authorizing a globally unprecedented 41,000-mile network of interstate highways [111], under the rubric of national self-defence, where the highways would act as vital escape valves from cities, should the country come under nuclear-bomb attack.[10] Other ambitious national highway projects ensued: South Africa's National Routes (1970s), Brazil's Trans-Amazonian Highway (1972) and India's Golden Quadrilateral project (1999). But perhaps nothing compares in scale to China's highway-building activity – over the past 30 years, as it has attempted to assert itself as one of the world's leading nations, it has built an entire network from scratch, totalling 136,500 kilometres, the largest in the world.

In all of these cases the same political argument has been made: that in order to modernize a country, it needs roads that will connect it – and preferably high-speed ones designed for cars. A mobile nation is a strong nation; an automobile nation even stronger.

← 111 The National System of Interstate and Defense Highways, published by the American Automobile Association, June 1958. The scheme was partly designed to connect the country by automobile, but also to provide conduits for the evacuation of people from cities should they come under nuclear attack

↓ 112 The first Ford Motor Company factory in Turkey, which opened in the Tophane district of Istanbul in 1925

Putting People Behind the Wheel

Simply connecting places with roads, though, was not good enough; these roads needed to be populated with automobiles, affordable to the masses. At first it looked as if Henry Ford was going to be the man to bring cheap cars to the world. With the wild success of the Model T in America, opening up car ownership to vast swathes of working- and middle-class Americans, Ford soon set his sights on conquering other foreign markets. In the brief period between his first operations in America and the onset of the Great Depression, he set about building a global empire of production, establishing assembly plants in Canada (1905); England (1911); France (1913); Argentina (1916); Denmark (1919); Brazil (1920); Belgium (1922); Chile and South Africa (1924); Turkey (1925) [112]; and Germany, India and Malaya (1926).[11]

Car manufacturers in Europe were especially keen to learn from Ford's manufacturing, so that they too might be able to scale up production and create a car for the masses, but also to fend off competition from Ford himself. A pilgrimage to Highland Park, Ford's main production site in Detroit, became de rigueur. Giovanni Agnelli of Fiat made a visit in 1912, as did André Citroën, and then another in 1923. Robert Peugeot's engineers also inspected the plant and its production techniques. Soon their European factories would take on new shape. André Citroën first introduced the moving assembly

line in 1919, and after a third visit to Detroit in 1931, he completely reorganized his factory once more along Fordist principles. Louis Renault, in 1928, with the chance to build a new factory on Île Seguin in the River Seine, also used the opportunity to institute a moving assembly line.[12]

Despite such European enthusiasm, none of the manufacturers during this period managed to scale up production anywhere close to the same degree as Ford. In France, Citroën's early attempts to replicate Ford nearly bankrupted him. In the UK, car models like the Austin Seven and the Morris Minor – directly inspired by the Model T – had become popular economy cars in the 1920s, but the transition to true mass production seemed littered with obstacles. William Morris, commenting on the state of manufacturing in the UK at this time, said: 'so far mass production has meant merely *mess* production when applied to motor cars in this country'.[13] A true people's car in Europe remained elusive.

Aside from carmakers, Ford had found himself another admirer on the Continent: Adolf Hitler. A framed portrait of Ford could be found in Hitler's private office as early as 1922, during the period when he was establishing the National Socialist Party.[14] Hitler specifically found a useful American ally in Ford's anti-Semitic tirades published in *The International Jew*, a publication that Ford himself produced and distributed, and which stoked fear and hatred of the Jewish people. But it wasn't just their shared anti-Semitism that drew Hitler to Ford; it was also a belief that mass production was the key to unlocking Germany's economic might. Hitler desperately wanted to reinvigorate the Germany car industry, which had stalled with the onset of the Depression, and the key, he believed, was the introduction of a cheap and affordable car for the masses – a people's car, or *Volkswagen*.

His plans began with a radio. From May to August 1933, under the eye of propaganda minister Joseph Goebbels, a consortium was established to produce a cheap and affordable radio, called the *Volksempfänger*, or people's receiver [113].[15] The radio was a massive success, bringing what was once a luxury commodity into the homes of millions of working-class Germans – and, most crucially, allowing for propaganda to be transmitted on a mass scale directly to individuals. It also emboldened Hitler to the idea that the state could play a leading role in stimulating national products, such as cars. In 1934 he attended the Berlin Auto Show, where he implored car manufacturers to emulate the success of the radio by making a car truly for the people. One year later this brief was formalized into a stringent set of demands – that the car should accommodate four to five people, cost less than 1,000 Reichsmarks, travel at a top speed of 80 km/h and consume between four and five litres of fuel over 100 kilometres.[16] The German automotive lobby responded by hiring Ferdinand Porsche, an Austrian engineer whose consultancy was based in Stuttgart. While the lobby had hoped that his engineering studies would prove the brief unfeasible, Porsche responded enthusiastically to the opportunity, quickly winning him the favour

→ **113** The *Volksempfänger* radio was the result of a Nazi campaign to create a people's radio that would be affordable for every household – and crucially would enable the dissemination of propaganda. It was held up as an example for the auto industry to follow, which ultimately resulted in the creation of the Volkswagen

of the Führer. On 11 July 1936 Porsche secretly drove two prototypes up to Hitler's mountain retreat, Obersalzberg, to show off his new creation to some of the top-ranking Nazi officials. Having won the confidence of Hitler and the party, Porsche was given the green light to dedicate the following years to refining the design, concerned now with the central question of how to scale up production.

Meanwhile, a few other measures were introduced to promote and facilitate future sales of the car. First, it was incorporated into a larger propaganda campaign and programme called *Kraft durch Freude* (KdF, meaning 'strength through joy') [114]. KdF was set up to help make leisure activities more available to the working and middle classes. It published posters advertising the pleasures of forest hikes, evenings at the theatre and road trips. KdF also sponsored low-cost holidays and outings, turning itself into the country's largest tourism agency. On 26 May 1938, when Hitler opened the impressively scaled new factory in Fallersleben that was to produce Porsche's creation, he christened the car the *KdF-Wagen*, recognizing the significant role

that the car would play in bringing the idea of the leisurely road trip to the people.[17] KdF followed through with an aggressive promotional campaign, issuing a postage stamp featuring the car and posters depicting happy couples racing through the countryside, while currying the favour of journalists through exclusive test drives. As credit was not a popular concept in Germany at the time, the party also had to invent a special savings plan to put the car within reach of many potential buyers. The *KdF-Sparkarte* was introduced as a scheme in which Germans would contribute five Reichsmarks per week, until they had accumulated the full 990 Reichsmarks purchase price.

All of this halted, of course, with the onset of the war and the conversion of the Fallersleben production site into a space for making military goods. As such, prior to the collapse of the Third Reich, only 630 cars were ultimately made,[18] hardly the mass-produced national car that Hitler had desired. But the case of Nazi Germany showed just how potent a symbol the automobile had become to the state of a nation. Through the combination of its ambitious *Autobahnen*, its national car and its leisure programme, the party shaped an image of the nation in which progress was measured in direct relation to the automobile [115]. This idea of the motorization of society became a benchmark for national development and was an example followed by many other countries after the war.

Road Tripping

Meanwhile the idea of the road trip found other unique manifestations. In each instance, these projects would show how the automobile was becoming a powerful tool for stimulating a tourist economy, which in turn helped to shape the cultural image of a country, through the codifying and listing of popular sites and attractions.

In France the tyre manufacturer Michelin realized early on that it could drum up enthusiasm for motoring by publishing small travel guides, which provided little city maps, as well as listings of hotels, restaurants and, most crucially, places one could get one's car repaired [117]. The first guide was published during the Paris *Exposition Universelle* in August 1900, taking advantage of the throngs of visitors descending on Paris. Although just 2,897 vehicles were registered in France at the time, 35,000 copies of the guide were printed,[19] signalling the Michelin brothers' belief that there was a severely untapped market for driving – and thus for tyres. The guides also created a new visual language, using small pictograms to denote quickly useful information: for instance, a train symbol meant that, should your car break down, you would be able to travel home by train; another symbol indicated whether a hotel had a darkroom, so that budding photographers could hurry back to develop their collodion plates. Beginning in 1904, Michelin began to expand its geographical scope, publishing first a guide to Belgium and subsequently to Algeria and Tunisia (1907); the Alps and the Rhine (1908); Germany, Spain and Portugal (1910); and Ireland and the British Isles (1911).[20]

Through these diminutive books Michelin was facilitating a whole new way of seeing and conceiving geography, while also playing a powerful role in defining the cultural attractions and the worth of a country. During the Second World War these books were even turned into a powerful weapon of navigation – while Michelin stopped publication at this time, a reprint of the 1939 edition was made so that American soldiers could better navigate French soil following D-Day.[21] Meanwhile, the competitor to Michelin's guides, Baedeker, was supposedly used to identify the British cities targeted for a series of bombing raids. Dubbed 'the Baedeker Blitz', the attacks were specifically designed to destroy Britain's cultural assets and thus demoralize the population [116].

Food – and, more specifically, culinary tourism – became closely associated with motoring and was used as a tool early on, to entice

↓ 116 Scenes of destruction following an air-raid attack on Norwich, as part of the German Luftwaffe's so-called 'Baedeker Blitz'

→ 117 'What Michelin did for tourism' poster, Michelin advertising extract from the magazine *La Petite Illustration*, by E.L. Cousyn, 24 August 1912, promoting its guides, maps, travel planning services and road signs. Collection du Patrimoine Historique Michelin, © Michelin

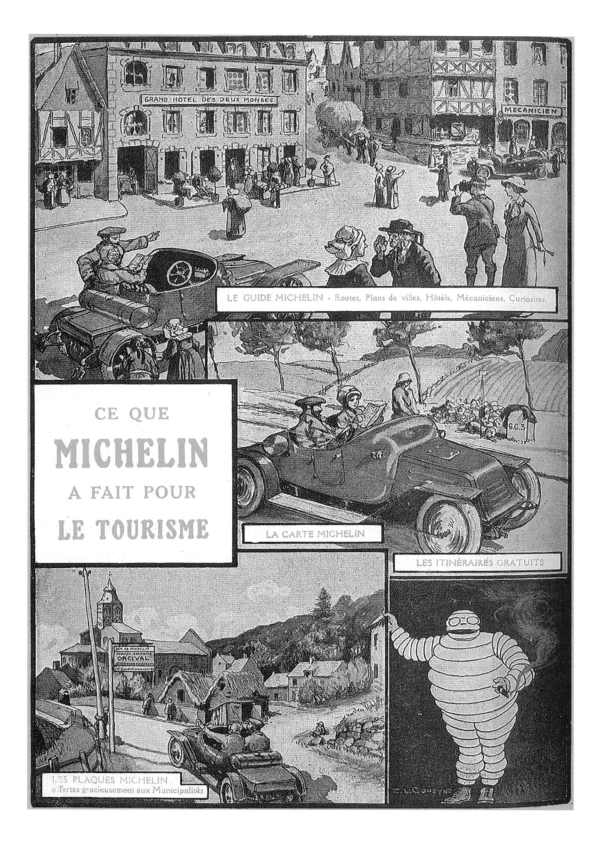

CE QUE
MICHELIN
A FAIT POUR
LE TOURISME

LE GUIDE MICHELIN - Routes, Plans de villes, Hôtels, Mécaniciens, Curiosités.

LA CARTE MICHELIN

LES ITINÉRAIRES GRATUITS

LES PLAQUES MICHELIN. offertes gracieusement aux Municipalités

people to drive. The Michelin rating system is perhaps the most enduring example. A star system was established at first as a simple way to rate hotels, but by the 1920s the guide had expanded to include restaurants of considerable quality, marked by a single star. By the 1930s the guide was awarding a fortunate few restaurants two and three stars, creating an immediate surge in demand for reservations and launching chefs and their enterprises to gastronomical fame.[22] The three-star system has remained ever since, in many respects outstripping in importance the original function of the guide and entrenching France's identity as a culinary leader.

Elsewhere, the combination of food and motoring took on spectacular new architectural forms. In Italy, during the economic miracle of the post-war years, ambitious new highways were spreading across the country, while the mass-produced Fiat 500s and 600s filled up the roads. A culture of motoring had opened up to the middle classes, who were now keen to partake in motorized exploration of the country and countryside [118]. The Pavesi-run Autogrill and its competitor, Mottagrill – early observers of this growing economy – responded by commissioning adventurous new architectural constructions to sit alongside the motorways, housing multifunctional service centres, with food positioned as the central attraction. The Autogrills served up a new kind of Italian 'fast food' modelled on American rest stops, to reflect the novel speed and

↓ 118 A 1960s Fiat 600 Multipla brochure promoting the car's use for sightseeing and road trips

→ 119 Render and elevation of the Autogrill Motta in Limena, by Pier Luigi Nervi and Melchiorre Bega, 1962

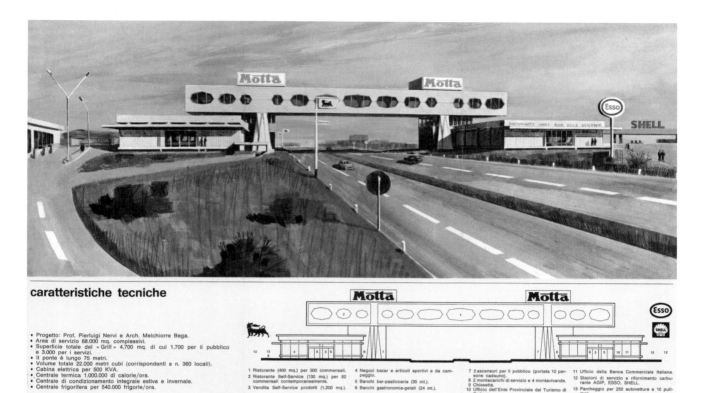

caratteristiche tecniche

- Progetto: Prof. Pierluigi Nervi e Arch. Melchiorre Bega.
- Area di servizio 68.000 mq. complessivi.
- Superficie totale del « Grill » 4.700 mq. di cui 1.700 per il pubblico e 3.000 per i servizi.
- Il ponte è lungo 75 metri.
- Volume totale 22.000 metri cubi (corrispondenti a n. 360 locali).
- Cabina elettrica per 500 KVA.
- Centrale termica 1.000.000 di calorie/ora.
- Centrale di condizionamento integrale estiva e invernale.
- Centrale frigorifera per 540.000 frigorie/ora.

1 Ristorante (400 mq.) per 300 commensali.
2 Ristorante Self-Service (130 mq.) per 82 commensali contemporaneamente.
3 Vendita Self-Service prodotti (1.200 mq.).
4 Negozi bazar e articoli sportivi e da campeggio.
5 Banchi bar-pasticceria (35 mt.).
6 Banchi gastronomia-gelati (24 mt.).
7 2 ascensori per il pubblico (portata 10 persone cadauno).
8 2 montacarichi di servizio e 4 montavivande.
9 Chiesetta.
10 Ufficio dell'Ente Provinciale del Turismo di Padova.
11 Ufficio della Banca Commerciale Italiana.
12 Stazioni di servizio e rifornimento carburante AGIP, ESSO, SHELL.
13 Parcheggio per 250 autovetture e 10 pullman.

culture of the highway, while Mottagrill exploited the demand for gastronomical tourism and the country's regional differences, to entice people to come and enjoy local flavours and specialities.[23] But while food was the lure, the architecture served as the real attraction. Recognizing that the speed provided by these new roads necessitated more bombastic architecture to catch the eye, Pavesi hired architect Angelo Bianchetti, who had cut his teeth earlier in his career designing trade-fair advertising pavilions. In 1959 he designed the first bridged Autogrill, located in Fiorenzuola d'Arda, between Parma and Piacenza, a new typology in which the building straddled the entire motorway.[24] Not only advantageous for being accessible from both sides of the motorway, but it also served as an unmissable icon for the passing motorist, and a novelty to be discovered and experienced from the inside. The glass facade of the restaurant enabled visitors to experience the thrill of the highway and become a spectator of their own speed, as they feasted on contemporary interpretations of a new 'fast' Italian cuisine. Motta quickly followed suit, hiring famed engineer Pier Luigi Nervi and architect Melchiorre Bega to build their own version of a bridged autogrill just outside Bologna. Over the following decade more than a dozen of these buildings were constructed across the country, becoming a visual symbol of modern Italy [119].

The Globalized Object

By the 1950s the car as a tool for economic stimulus and a measure of economic might had become well recognized. It was a sentiment perfectly encapsulated in the statement 'What is good for the country is good for General Motors, and what is good for General Motors is good for the country', a much-circulated misquote from General Motors' CEO, Charles Wilson, during his confirmation hearing to become Secretary of Defense in 1953. But a visual language also emerged, equating certain looks with specific national traits. As Steven Parissien writes of the period, 'American autos were seen as brash, large and loud; Italian as racy and flashy; British as well built and sedate.'[25]

The irony of all of this is that, since their debut, automobiles have been a globalizing force *par excellence*. As we have seen earlier in this chapter, Ford began expanding his empire of factories around the world almost from the start. And today, for any car company to be successful, it needs to establish a carefully coordinated system of global supply-chain management, with parts manufactured, shipped and assembled in disparate places around the world rather than in one country, let alone one factory site. Even the ultimate national car by design, the Volkswagen, found its real success after the war not in Germany, but in America, inhabiting its own unique identity stripped of its previous nationalistic connotations (thanks to a clever advertising campaign, pp. 109–10).[26]

One car in particular highlights perfectly well the perversity of this confused and contradictory condition: how an object can inhabit both a national identity and a globalized reality at the same time. That is Iran's national car, the Paykan. In the 1960s the pro-American shah, Mohammad Reza Pahlavi, believed – like many leaders of developing nations – that motorizing Iranian society was one of the keys to its modernization. He set about ambitious road-building projects, yet a domestic auto industry remained elusive. Seeing a massive opportunity to introduce affordable cars to the emerging Iranian middle class, Mahmoud Khayami, co-founder of the company Iran National, struck a deal with the British car manufacturer Rootes, to license complete knockdown kits of their Hillman Hunter. By 1967 these kits were being packaged in the UK and then shipped to Iran, where they were assembled under the name Paykan ('arrow' in Farsi).[27] For its third anniversary, the company commissioned film-maker Kamran Shirdel to produce a commercial celebrating three years of success. The car appeared to come out of a giant birthday cake, as happy couples danced around it. Not wanting to mimic the well-known American 'Happy Birthday' song, Shirdel had a new birthday song commissioned, sung in Farsi. Signalling just how important this car was in the popular imagination, the song went on to become the standard song that all Iranians sang (and continue to sing today) at birthday parties, even if the advert has long been forgotten [120].

The Paykan became a major success and, remarkably, survived the violent upheaval of the 1979 Islamic Revolution. If at first the

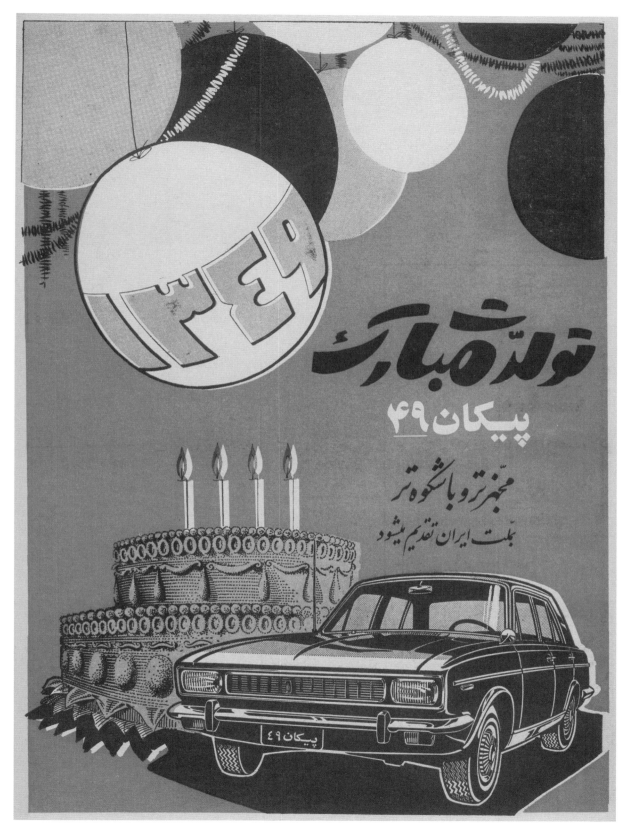

car stood for an American-loving decadence and the embrace of unbridled modernity, as typified by the period under the shah, then following the regime change it became lionized as a national war hero, playing a crucial role in the Iranian fight against Iraq during their eight-year war [121]. The Paykan was by far the most popular car on Iranian roads, and its dominance lasted until production finally ceased in 2005. As a result, its image is deeply engrained in the national imaginary of its citizens as being quintessentially Iranian. This is ironic, as it came originally from Britain, from a company (Rootes) bought out by an American company (Chrysler) [122], and a decade later by the French (Peugeot).

This changing of hands and blurring of national lines is entirely consistent with broader automotive history. Look at a timeline of car companies from the beginning of the twentieth century to the present day and you will see a great confluence. An industry that once comprised hundreds of independent manufacturers has, over time, merged into a handful of massive multinational corporations. General Motors pioneered the trend early on, buying up rival car companies in Michigan to become the largest car company in America. But others have followed, blurring lines of nationality and national affiliation. The Volkswagen Group's impressive portfolio encompasses the Audi, Bentley, Bugatti, Lamborghini, Porsche, SEAT and Škoda brands. Given the strong national and local identities that so many of these brands have cultivated over the years, the mergers can make for jarring contrasts. The diminutive Fiat and the brash Chrysler make for strange bedfellows, following their merger after the 2008 financial crisis. And sitting in the back of an iconic black cab while driving through London, you would be forgiven for not suspecting that it is in fact owned, as of 2013, by the Chinese company Geely. The cars that were once used to build the image of rising states are now members of an increasingly fluid, murky and transnational amalgamation of objects. Used as tools to shape nations, cars have been quietly working to unshape them as well.

↑ **121** Street celebrations on
the day Mohammad Reza
Pahlavi, the Shah of Iran,
left the country, with a
Paykan in the foreground,
photograph by Bahman Jalali,
16 January 1979

→ **122** A 1968 Chrysler advertisement showing the diversity of its product range at the time. The Sunbeam Arrow in the bottom right corner was alternatively badged as the Paykan when it entered the Iranian market

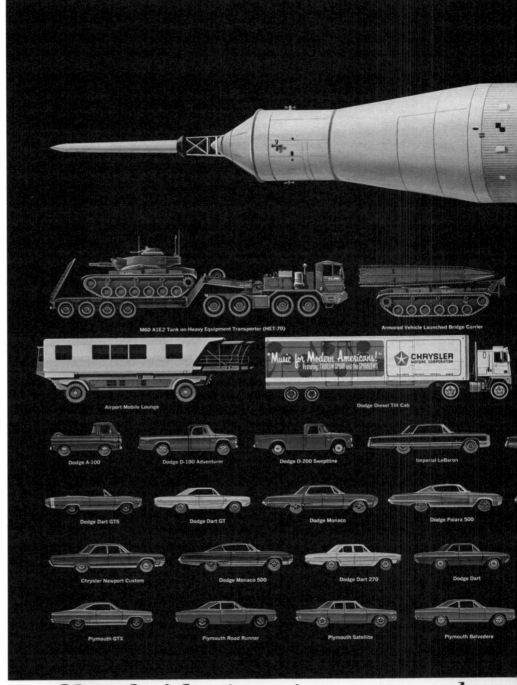

M60 A1E2 Tank on Heavy Equipment Transporter (HET-70)

Armored Vehicle Launched Bridge Carrier

Airport Mobile Lounge

Dodge Diesel Tilt Cab

Dodge A-100

Dodge D-100 Adventurer

Dodge D-200 Sweptline

Imperial LeBaron

Dodge Dart GTS

Dodge Dart GT

Dodge Monaco

Dodge Polara 500

Chrysler Newport Custom

Dodge Monaco 500

Dodge Dart 270

Dodge Dart

Plymouth GTX

Plymouth Road Runner

Plymouth Satellite

Plymouth Belvedere

Chrysler's business is to get you where

At Chrysler Corporation, we make things that move.

They splash across lakes. They burrow through marsh mud. They haul pig iron and they haul petunias. They even go so far as to take people out of this world.

You'll find us moving in 130 different countries—with everything from irrigation pumps that m[...] water to Airtem[...] housefuls of coo[...]

And the ni[...]

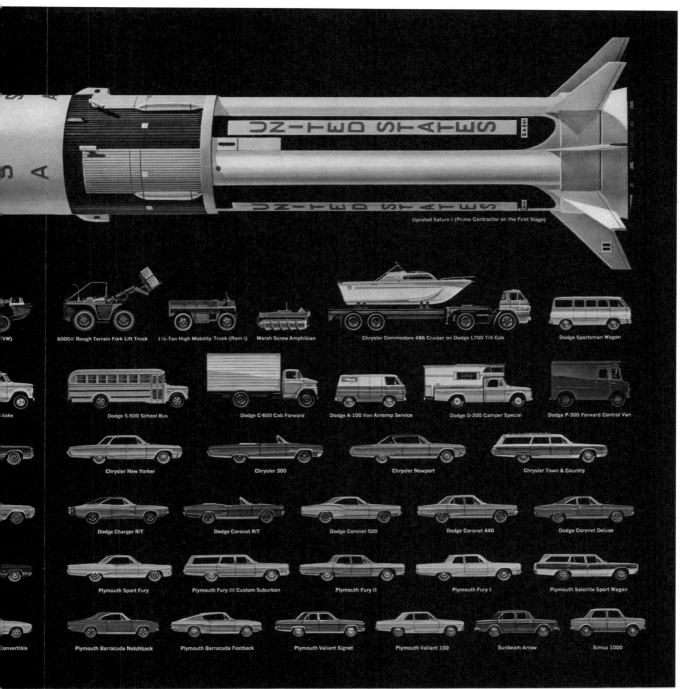

Uprated Saturn I (Prime Contractor on the First Stage)

6000# Rough Terrain Fork Lift Truck	1¼-Ton High Mobility Truck (Ram I)	Marsh Screw Amphibian	Chrysler Commodore 486 Cruiser on Dodge L700 Tilt Cab		Dodge Sportsman Wagon
Dodge S-500 School Bus	Dodge C-600 Cab Forward	Dodge A-100 Van Airtemp Service	Dodge D-200 Camper Special	Dodge P-300 Forward Control Van	
Chrysler New Yorker	Chrysler 300	Chrysler Newport	Chrysler Town & Country		
Dodge Charger R/T	Dodge Coronet R/T	Dodge Coronet 500	Dodge Coronet 440	Dodge Coronet Deluxe	
Plymouth Sport Fury	Plymouth Fury III Custom Suburban	Plymouth Fury II	Plymouth Fury I	Plymouth Satellite Sport Wagon	
Plymouth Barracuda Notchback	Plymouth Barracuda Fastback	Plymouth Valiant Signet	Plymouth Valiant 100	Sunbeam Arrow	Simca 1000

...ant to go—even if it's 238,000 miles straight up.

...lons of ...at move ...hrysler

Corporation engineering is, we also make sure everything *keeps* moving. Mile after mile. Year after year. Maybe that's one reason why we've moved up to being the fifth

largest industrial corporation in America.
Plymouth · Dodge · Chrysler · Imperial · Dodge Trucks Simca · Sunbeam · Airtemp · Cycleweld · Oilite · Mopar Parts Division · Marine and Industrial Products · Defense and Space Products · Leasing · Financing · Insurance

CHRYSLER CORPORATION

POLITICAL SYMBOLISM AND THE NYAYO PIONEER CAR

Nanjala Nyabola

In 1982 members of Kenya's air force attempted a *coup d'état* to overthrow Daniel Toroitich arap Moi, who at the time had been in office for just four years. The effort was quickly quashed, but heralded an unprecedented era of authoritarianism in the country. Many of the superseding patterns of repression echoed the practices of the British colonial administration, from which Kenya had never fully decoupled. The capital city, Nairobi, was littered with heavily protected sites where torture was routinely used to extract information from suspected dissidents. Seemingly every day thousands of soldiers, university students, politicians and ordinary people who were in the wrong place at the wrong time would be siphoned into these spaces and often never heard from again.

Six years later, on 29 February 1988, Moi was selected unopposed as president by the ruling party, beginning his fourth term. But by this time turning the screw on suspected dissidents was having a counterproductive effect. The more the government arrested them, the more they emerged from the woodwork, often organizing around the label 'Mwakenya' – a banned group advocating multiparty democracy, which the government accused of treason. Between March 1986 and November 1988 there were 95 political trials and 100 official political prisoners in Kenya, most of the cases concerning the creation and distribution of Mwakenya materials.

At the same time international financial organizations that had been key contributors to Kenya's economy (as part of a broader western resistance to communism in Africa) were also disenchanted with the country. Throughout the 1980s Kenya had been non-compliant with many of the demands by the International Monetary Fund (IMF) for austerity, finding a measure of quarter from consequences by enthusiastically embracing capitalism. When it became clear that communism would not survive, that measure of protection was lost, and by 1986 structural adjustment programmes – broad austerity and trade liberalization initiatives proposed by the IMF – were part of the country's economic planning papers.

Between the fear and the economic upheaval, Kenya threatened to career off the political rails, and Moi desperately needed a win. So, on 28 February 1990, almost two years to the day since he began his fourth term, and in the presence of thousands at the Kasarani National Stadium, with his ubiquitous *rungu* – a small wooden baton with a rounded head that Moi wielded as a physical manifestation of his authority – in hand, Moi hopped into the driver's seat of a small, boxy saloon car and drove a single 400-metre lap around the race track. One small lap for the president, but one giant leap for the country, he probably hoped. But was it?

L'état c'est Moi

The boxy off-white saloon car with the dark-brown interior that Moi drove around the Kasarani track, while waving his *rungu* in a familiar gesture of paternalistic pandering, was the Nyayo Pioneer [123]: a car built from a joint initiative between the University of Nairobi, the Kenya Railways Corporation and other technical partners. Its cream exterior glimmered and its lines were clean, but its origin story was murky, steeped in the bloody waters of Kenya's complex political history. From inception through to its launch, the car was an overt ploy to rewrite Kenya's increasingly dark and ugly political trajectory – one of Moi's many attempts to mask his increasing authoritarianism with grand, ill-conceived white-elephant projects.

The word 'Nyayo' itself is a sobriquet for Moi, and the proliferation of Nyayo-named projects and initiatives during the 1980s was emblematic of the collapsed boundaries between the state and the head of state. Before the Nyayo Pioneer project, there were the Nyayo Bus Services and the Nyayo Tea Zones (special trade zones for the growth and processing of tea). In Swahili, 'Nyayo' means footsteps, because Moi promised, on his inauguration, that he would follow in the footsteps of Kenya's founding president, Jomo Kenyatta. He named his national philosophy 'Nyayoism', and every Friday morning, before receiving their daily ration of free Nyayo Milk, school children across the country participated in a

↑ **123** In 1986 President of Kenya, Daniel Toroitich arap Moi initiated a project in collaboration with the University of Nairobi to manufacture Kenyan cars, and the result was the Nyayo Pioneer, 1990

mandatory recitation of the loyalty pledge, in part swearing fealty to 'the Nyayo philosophy of peace, love and unity'.

In the global sense, Moi was certainly not unique. The 1980s represented the golden age of the 'Father of the Nation' trope – a system or style of governance in which a political leader places himself at the heart of government, governance and every routine function of society. The trope is often framed by an eponymous ideology, based on internally consistent but externally irrational and ill-considered premises. Arguably, the Father of the Nation trope has existed in some form everywhere in the world, but it was communist regimes in the USSR and China that gave African countries like Kenya a blueprint. Stalinism and Maoism provided leaders like Moi with a template with

which to collapse the state into the leader, and to consolidate power around the individual even while communism per se remained mistrusted.

The Pioneer car was one of a number of Nyayoist nationalist projects intended to remind Kenyans that the state (in the form of the head of state) was still capable of magnificence. These projects invited increasingly restless publics to defer their anger while waiting for the president to deliver on the rest of his political promises. While their proponents may genuinely have believed in them, the effect of these nationalist schemes is best described by the French phrase '*il bouge*' (he moves): periodic reminders from on high that the specific ruler, and therefore the state itself, was active, and that progress was imminent if the citizens would just be patient.

Grand Designs, Made in Kenya

At the 1986 University of Nairobi graduation ceremony, Moi challenged engineering students at the university to build a national car for Kenya, 'no matter how ugly or how slow'. Superficially, this was a harmless invitation for the best university in the country to make itself more relevant to the socioeconomic life of the nation. But there was more to the story, and the urgency with which the university acted on the directive speaks of the climate of fear in the country at the time.

One of the institutions that came out strongly in support of the 1982 attempted coup was the University of Nairobi, where liberal-minded lecturers and radical student leaders had long resisted authoritarianism under Moi and his predecessor, Jomo Kenyatta. As part of the broader effort to undercut political organization at the university, the institution was closed for 14 months and was fundamentally restructured. Security agents infiltrated the lecture halls, staffrooms and student groups, arbitrarily arresting and detaining students and staff members who were deemed too radical. Indeed, after 1982 students needed special clearance from the Office of the President to hold any kind of meeting, including religious services. Still the strikes and resistance continued well into the 1990s, as each incoming generation found the environment at the campus increasingly intolerable.

In February 1985 the university administration violently broke up a protest prayer meeting called by a student organization, and one student was shot and killed in the resulting mêlée. At the same time the 1985 University of Nairobi Act notably omitted a provision on academic freedom that had been in the original founding Act of 1962. The status of the university as a place of open academic enquiry had already been under threat, but increased repression – flavoured by a large helping of funding cuts demanded by international financial organizations – crippled the institution altogether. 'Kenya has the knowledge and manpower to spar with the best of the world in manufacturing', declared Moi at the 1986 graduation ceremony, even as his administration was actively undermining the foundational principles of such creativity and innovation.[1]

The Nyayo Pioneer car was thus a little different from other Nyayoist projects. Yes, it was another effort at appeasement and misdirection but, like a parent redirecting the energies of a hyperactive teenager, it was also an effort to redirect the frenetic energy at the campus. For the university administration it represented a chance at redemption after many public failures to control political resistance on campus. And for Moi it was a chance to affirm to Kenyans and outsiders that Nyayo and Nyayoism – and, by inference, Kenya itself – moves.

Four years after Moi's initial challenge, the first prototypes of the Pioneer car were launched. Given that the one functional car of the five prototypes struggled even to complete the 400-metre circuit, there is some speculation that the launch was rushed, in the shadow of one of the darkest events in Kenya's political history: the gruesome assassination of the former Minister of Finance, Robert Ouko. In his autobiography former US Ambassador to Kenya, Smith Hempstone, alleges that Moi himself pulled the trigger on Ouko.[2] But Hempstone's book remains banned in Kenya, so most Kenyans had nothing more than the official narrative to hold on to: on 12 February 1990 Ouko shot himself in the head and then took off his clothes, folded them neatly in a pile, poured acid on himself and set himself on fire.

The deeply suspicious nature of this account perhaps explains why, in his speech launching the Pioneer car, Moi took time to caution 'those who take advantage of national tragedies to concoct malicious falsehoods'.[3] Indeed, the front page of the *Daily Nation* was shared by the announcement of Kenya's great leap forward and an article in which Moi warned threateningly of the danger of rumours. To the university community who were present at the launch, he asserted that universities in Kenya flourished only because he loved 'the youth', and that lecturers in Kenya were lucky compared to their counterparts in other countries. These stern and perhaps ominous words reasserted the state and the head of state as the paramount chief and supreme energy, from which all opportunity and progress in the country flowed.

Technically, the Nyayo Pioneer car wasn't terrible – the prototypes were arguably a good start for a country with no tradition of automotive engineering. The saloon model that Moi drove around the Kasarani track had an engine capacity of 1200cc and a maximum speed of 160 km/h, with a cruising speed of 120 km/h. Its interior was chocolate-brown with dark accents, acquiescing to the simplicity that Moi had urged when commissioning the car. There is truly nothing remarkable about the design – it looks much like any other saloon car of its time, except perhaps with a higher-than-average chassis in order to accommodate Kenya's rough roads. While the boxy shape seems dated today, it was typical of its era, especially because the prototype was cobbled together from the parts of various vehicles assembled in Kenya by international manufacturers. And while the car struggled to limp over the 400-metre mark, it was priced at just 160,000 Kenya shillings – well below the usual price range of a brand-new saloon car during that period.

In hindsight, it is hard to gauge the actual reaction to the Pioneer car. There was, and remains, an element of excitement over the car itself and what it could represent. Similarly, the media at the time gushed about this giant leap forward for Kenyan engineering and development in general, but in the context of a media captured by an authoritarian state, it is difficult to discern what was genuine excitement and what was a terrified industry parroting the enthusiasms of the state. What is unambiguous, though, is that the state had managed to redirect the interests and passions of the entire nation for several moments, perhaps deflecting the chaos that should have superseded the assassination of Robert Ouko.

Political Contexts and Meaning-Making

Political contexts can impart objects with meaning that is far greater than the intention of those who created them. And the creators often lose control of their symbolism as soon as the object enters common or public use, particularly in authoritarian contexts where controversial opinions may not be overtly expressed or consciously acted upon. Thus the Nyayo Pioneer car has come to symbolize Kenya's stalled transition to democracy: a crisis of political imagination and the cyclical ideological masturbation of its political class.

Eventually the Pioneer project would consume at least 750 million Kenya shillings as it sputtered towards infamy, as one of the most notorious white-elephant projects of the Nyayo era. Only the five prototypes on display were ever built, and of those five, only the saloon car was ever driven. The bulk of this investment nominally went to the creation of manufacturing plants for the car, but by 1999 the one completed plant had choked under the strain of austerity and broader economic mismanagement. The Nyayo Manufacturing Company, founded to make the car, was one of the many state corporations privatised as part of the austerity push, re-emerging as the Numerical Machining Complex. Although it has never manufactured a car, the organization maintains an office in which it houses the five Pioneer prototypes.

In December 2002 Moi retired to a vast farm in his hometown of Baringo, but the old man maintains a firm grip over Kenyan politics. Every major politician in the country makes a publicized pilgrimage to Baringo before embarking on a national political career, leaving no doubt as to who still calls the shots in Kenya. Meanwhile, misguided nostalgia for Nyayoism remains the hallmark of Kenyan public policy. Predictable perhaps, considering that (with the notable exception of Raila Odinga, who was detained and tortured from 1982 to 1988) all of Kenya's current national political figures are Moi protégés.

Kenya, like Nyayoism and the Pioneer car itself, moves – *il bouge* – but in 2018 the race to govern Kenya remained a misguided race to complete the promise of Nyayoism. Since 2013 the National Youth Service (NYS), the promise of free milk for every school child, as well as numerous ill-advised infrastructure projects like the Lamu Coal Plant – all previously shelved by a state that could not afford

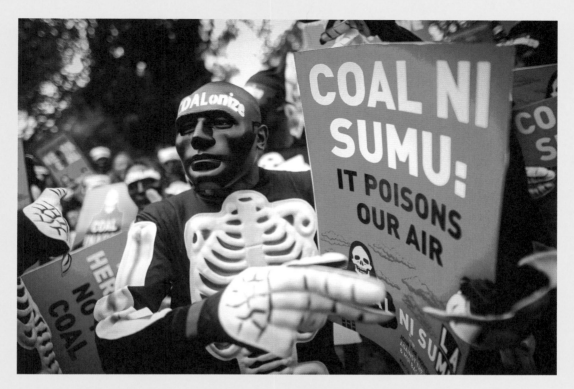

↑ **124** Kenyan activists protest against the Lamu Coal Plant in Nairobi, 5 June 2018

both the expense and the voracious appetites of Kenya's corrupt elite – have made a comeback [124]. The only difference is that this time the dominant ideology is profit: the projects are being reintroduced as profit-making opportunities for Kenya's connected few.

Unsurprisingly, therefore, in 2011 the Ministry of Industrialization trotted out the Nyayo Pioneer prototype at a trade fair, declaring the revival of the project as a public–private initiative. As a testament to the soundness of its design, during the event the car chugged along for several metres, even though it had not been driven in more than 20 years. But Kenya had changed considerably in the intervening period, and no Father of the Nation could order progress by decree. And so, echoing the country's stalled democratic promise, the Pioneer sputtered back to its storage facility, awaiting the next opportunity for nationalism on demand.

THE EXPRESS CITY

Brendan Cormier

CITY

The dramatic growth of car ownership over the past century coincided with one of the most profound sea changes in history: the rapid urbanization of cities and regions around the world. The growing public demand for cars required that cities be conceived and built in entirely new ways – architecture was reoriented primarily to serve the car, roads were vastly expanded and widened, and expressways were planned to criss-cross dense and sometimes implausible geographies. This new car-oriented landscape also changed the way we see, experience and inhabit the city – from the over-sized signage of city streets, to the increasing hours logged sitting idly in cars.

Such urban transformations have long captivated photographers, who have focused their lens on the dramatic consequences of our car-centric urbanization. This photo essay features the works of artists who reveal in different ways the realities and contradictions of the city built for the car.

↑ 125 *Nanpu Bridge Interchange, Shanghai, China*, photograph by Edward Burtynksy, 2004

Scale of Impact

Edward Burtynksy's work focuses on the large-scale impact of human industry, rendered into surreal large-format landscapes. His city photographs, like this one of the Nanpu Bridge Interchange in Shanghai, reveal the complex logistic infrastructure required to keep cars, trucks, planes and ships moving in urban environments.

↖ 126 *Hot Shot Eastbound
at Iaeger*, West Virginia,
photograph by O. Winston
Link, 1956

↑ 127 *On the Set of 'American
Graffiti', Filmed in and Around
the Bay Area*, photograph by
Dennis Stock, 1972

Buildings for Cars

To better accommodate the car in the fabric of the city, engineers,
designers and architects dreamed up entirely new types of buildings.
The drive-in cinema and the drive-in restaurant were two of the most
popular such inventions. In *Hot Shot Eastbound*, o. Winston Link
captures the spirit of a night out at the drive-in cinema in the 1950s,
while Dennis Stock's documentation of the sets of *American Graffiti*
(1973) help to highlight the nostalgia that the drive-in restaurant
has garnered over time.

Systems of Communication

Denise Scott Brown is an architect who co-authored the influential 1972 book *Learning from Las Vegas*. In it, she explores how the city's architecture of glitzy casinos acted as pseudo-billboards, attracting potential visitors from the adjacent busy highway. Her photographs reinforce this idea: cities have transformed into systems of oversized signs and graphics, to be seen and understood from the perspective of a fast-moving car.

↓ **128** *Pico Boulevard, Santa Monica*, photograph by Denise Scott Brown, 1966

→ **129** *Architettura Minore on The Strip, Las Vegas*, photograph by Denise Scott Brown, 1966

← **130 & 131** *66 Drives*,
photographs by Andrew Bush,
1989–97

↑ **132** *Toshi-e*, photograph by
Yutaka Takanashi, 1974

Occupying the City

As more and more people buy automobiles and commute to work by
car, more of our daily city life is spent couped up behind the wheel. It
has been estimated that Americans, for instance, spend an average
of 17,600 minutes driving each year. Andrew Bush documents the
almost mundane absurdity of this lost time, isolated yet public,
through a series of side on portraits of drivers in Los Angeles, while
Yutaka Takanashi captured early car commuting culture in Japan.

Monumental Highways

Highways have become mandatory infrastructure for any major city looking to efficiently move people around. In the dense and topographically varied cities of Italy, this has proven to be a substantial engineering challenge. In her photographs, Sue Barr captures the often-jarring juxtaposition of monumental highway construction above, with the more humble and diminutive city structures below, examples of which can be found across the country.

↙ → 133 & 134 *The Architecture of Transit*, photographs by Sue Barr, Italy, 2014

↖135 *Fragmented Cities*, Monterrey, Suburbia Mexicana, photograph by Alejandro Cartagena, 2006

↑ 136 *Fragmented Cities*, Santa Catarina, Suburbia Mexicana, photograph by Alejandro Cartagena, 2007

The Image of Sprawl

Cars and highways have fundamentally helped spur the development of suburbia – tracts of low-density single-family housing that have caused cities to sprawl deep into the countryside. Alejandro Cartagena's series *Suburbia Mexicana* shows how this process has recently played out on the periphery of Mexican cities. Similarly, Alex Maclean is interested in the traces that such development leaves on the fringes of American urban landscapes, whether that be speculative curvilinear roads and cul-de-sacs, or an isolated square of suburban housing.

↓ **137** *Desert Housing Block,*
Las Vegas, NV, photograph
by Alex Maclean, 2009

→ **138** *Desert Overlay,*
Meadview, Kingman North,
Arizona, USA, photograph by
Alex Maclean, 2009

THERE ARE NO CARS

Allison Arieff

IN WAKANDA

People can't get around conveniently because they are far away from everything.

André Gorz, *The Ideology of the Motor Car* (1973)

In continuing to look at the car as some magical conduit to a brighter future, we continue to ignore what the automobile has wrought. When we're so enamoured with the way technology might transform the car – and, by extension, our lives – we fail to explore adequately how getting rid of cars might transform how and where we live. We'd do well to heed André Gorz's exhortation to 'never make transportation an issue by itself'.[1]

What if we could rethink mobility to be not about the car, but about people? What if we thought less about technological innovation and more about connection and community, equity and access? Might it be possible to imagine a move away from petrol? From drivers? From cars?

As journalist Derek Thompson wrote in *The Atlantic*, 'I have seen the future of the car. It looks like a minivan.'[2] Thompson was writing about a driverless minivan, but still. It is truly surprising to observe in the twenty-first century that the car of the future – driverless, electric, flying or otherwise – looks like … a car. What is radically different now is that the means to make that car drive autonomously have been figured out. Companies testing autonomous cars have logged millions of driving miles (though not without a few high-profile crashes). Many experts, from architects to automobile executives, predict the ascendancy of the autonomous vehicle within three generations. Insurance companies are now preparing actuarial tables for them; in an unusual shift towards long-term planning, Ford, BMW, Audi, Mercedes-Benz and Nissan, among other car manufacturers, are seeing the writing on the wall and have developed working prototypes.

Yet these 'futuristic' cars are often weirdly anachronistic. The CV-1, a prototype electric car developed by the Russian arms manufacturer Kalashnikov, is presented in a render as floating in space, evoking memories of the space race of the 1960s.[3] Indeed, the CV-1 harks back to that era, taking its aesthetic cues from the Izh, a boxy Soviet hatchback first introduced in 1973. Meanwhile in Silicon Valley, Google's autonomous vehicle division, Waymo, went all in on the Chrysler Pacifica *minivan*, the most mundane car model imaginable. And upmarket carmaker Audi's rendering of its luxury AV features a bearded millennial in sweater and trousers reclining in the back seat, engaged in that most analogue of activities: reading a hardback novel.

As uninspiring as these cars are in their design, what is more egregious are the ways they increasingly chip away at social interaction. Russian inventor Semenov Dahir Kurmanbievich's Patent #2428364[4] is perhaps best described as 'extreme drive-through'[5]: the car isolates its driver from the world outside by design. A video illustration of Kurmanbievich's concept shows a man driving into a shopping centre, selecting his groceries from a conveyor belt while still in the front seat and, incongruously, presenting his credit card (no Apple Pay?) to a cashier, while his groceries somehow magically get into the boot. Flatpack furniture giant Ikea funds a think tank, SPACE10, which in 2018 prototyped seven public-service units[6] that could be summoned to any location via an app. Ikea's rationale was that these wheeled pods could deliver a suite of services to remote or underserved neighbourhoods: 'Healthcare on Wheels' would help doctors reach infirm patients who could not travel to clinics, while 'Farm on Wheels' might deliver fresh produce to areas without access to it. But what does the city look like, if everything is delivered directly to us? What does our community look like, if we never leave the house?

Indeed, prototype after prototype espouses a future where we never have to be bothered by other people; where we are, in fact, shielded from them. Consider the rendering of Volvo's 360c autonomous vehicle, featuring a person who appears almost hermetically sealed within it [139].[7] The 360 not only looks like a first-class aeroplane suite, but it also intends to replace one. Actually it could possibly replace a small apartment, albeit a luxury version: its interior has been designed with a fridge, bar, sink and foldout bed. If life is a highway, why not live on one? Making car travel so effortless feels like a gracious invitation to endless suburban sprawl.

If you can read your iPad, enjoy a cocktail or play a videogame while commuting, time spent in the car becomes leisure time – something desirable. Long commutes are no longer a disincentive.

With this sort of amenity-rich cocoon, there is a relentless focus on the object, absent of any context or community. The future of mobility is assumed to be car-dependent, while a vision based more on public transportation is thought to be old-fashioned. But isn't planning for the car the thing that is most out of date?

What is perhaps most notable about these renderings, sketches, videos and the like is what they leave out. We don't see pollution or smog, traffic jams or petrol stations, sprawling surface parking areas or car accidents. No people of colour. Or old people or the homeless. Indeed, these visions deliberately exclude anything that might be perceived as an obstacle or that seem outside a very narrow norm, from regulatory impediments to the inconvenience of other people. This characterizes the worldview of those who can only see a future that continues to be designed for the car. At its most extreme is perhaps Tesla's CEO, Elon Musk. His company might be pushing electric, but in the end it's still pushing the car:

> I think public transport is painful. It sucks. Why do you want to get on something with a lot of other people, that doesn't leave where you want it to leave, doesn't start where you want it to start, doesn't end where you want it to end? … It's a pain in the ass. That's why everyone doesn't like it. And there's like a bunch of random strangers, one of who might be a serial killer, OK, great. And so that's why people like individualized transport, that goes where you want, when you want.[8]

It's little wonder American billionaires focus their energies on private cars and/or away from public transit. There is no end to the public-transit workarounds that Musk has proposed, from digging personal car tunnels under Los Angeles to lauding the speed and superiority of the as-yet-unbuilt Hyperloop – a pneumatic tube that hurls passengers through a tube at 400 mph. Similarly the Koch Brothers, conservative oil barons in the USA, have been quietly paying for massive mobilization efforts to defund or derail public-transit projects across the country. Like Musk, the Koch Brothers see public options as antithetical to a free market – and, by extension, to liberty. As a spokeswoman from the Koch-funded advocacy group known as Americans for Prosperity put it, 'If someone has the freedom to go where they want, they're not going to choose public transit.'[9]

Among the many challenges in a capitalist society, it's hard to imagine a way to profit from a car-free future. Governments don't profit from providing public transportation to citizens. It's a service, a reflection of our commitment to the social good. Private transportation – cars – is a different animal: hundreds of industries profit from it, from oil companies to media conglomerates. A less car-dependent society necessitates a rethinking of this system.

In the meantime could this continual disdain for, and defunding of, public transit take us to something out of *Mad Max*, the classic film of 1979 in which survival is predicated on the ability to move – and move fast. In this dystopian scenario, the depletion of resources (natural and otherwise) has not only led to extreme hoarding, but has also utterly erased the public realm. There is no built environment *but* the car.

Mad Max's script drew heavily on the effects of the 1973 oil crisis on Australian motorists, its screenwriters basing their script on the thesis that 'people would do almost anything to keep vehicles moving and the assumption that nations would not consider the huge costs of providing infrastructure for alternative energy until it was too late'. In this, they were highly prescient: the world's collective apathy in the face of increasing temperatures, catastrophic weather events and climate refugees – not to mention the inevitable gas shortages, which might feel tame in comparison to those experienced by the *Mad Max* hooligans – is staggering.

Today 'doing just about anything is an environmental question', writes philosopher of the Anthropocene, Timothy Morton. He continues:

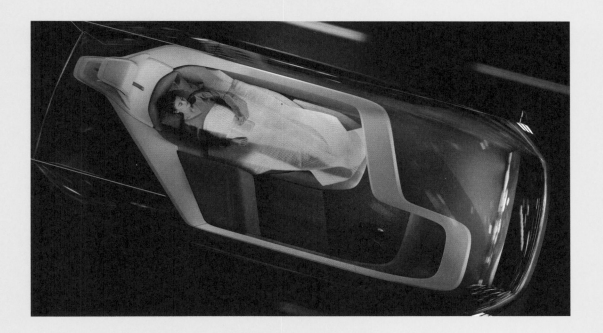

↑ **139** The Volvo 360c Interior Sleeping concept car is an autonomous, electric vehicle designed to allow passengers to sleep, work or relax while they travel, 2018

That wasn't true 60 years ago – or at least people weren't aware that it was true. Tragically, it is only by despoiling the planet that we have realised just how much a part of it we are ... There you are, turning the ignition of your car. And it creeps up on you.

Every time you fire up your engine you don't *mean* to harm the Earth, 'let alone cause the Sixth Mass Extinction Event',[10] but 'harm to Earth is precisely what is happening.'[11]

Automobiles harm the earth. Cars are the single largest contributor to air pollution in the US and – short of a full-scale switch to electric, which seems unlikely, given the troubled history of electric (and healthy cadre of lobbyists) – will continue to be. Yet almost all urban land has been, and still is, designed to accommodate cars. But what if it no longer was? What if we didn't let the car determine the design of our cities and the pattern of our daily lives?

From Ford's Model T to Waldo Waterman's 'Whatsit' (a flying car prototype designed in 1932), and from the quintessential nuclear family in *The Jetsons* to Google's twenty-first-century 'Opener'[12] (a flying car concept now described as evTOL, or 'electric vertical takeoff and landing'), the car has been seen as enabling the path to progress. Futurists, technologists, engineers, film directors, science-fiction writers, automobile execs, transportation planners, venture capitalists – almost all are white males. And this is one of the primary reasons our collective visions of the 'future' feel so limited. They're not collective at all, because so few have been empowered to imagine the future at all, let alone unleash their vision of it on a broader public.

What if we let more people imagine the future, concerning how we move about? What if more people could have a say or offer a vision? And what if instead of always focusing on how to get somewhere else, we focused on *where* we are?

Cities and towns would need to rethink land use, and their housing and transportation policies. Governments – local, national and regional – would need to reprioritize their investments, from roads and bridges to transit and walkability. But above and beyond these pragmatic considerations, such shifts would require a rethinking of the American Dream (and the Quarter-Acre Dream and similar variants) – of the notion that success is embodied only in the form of a single-family home and a two-car garage. It would be a massive cultural and behavioural shift to radically recast the automobile as a symbol not of freedom, but of restriction.

We need to show how liberating *not* having a car can be. Architects, planners and developers of both cities and suburbia have advocated, and continue to advocate, what the real-estate development company Gerding Edlen calls '20-minute living' or 'being able to do all of the necessary and enjoyable things that make life great within 20 minutes of your home ... twenty minutes on foot is ideal but 20 minutes by transit, bike or even auto is a reasonable goal'.[13] There are practical ways to achieve this goal, from creating a retail main street for every neighbourhood, to providing quick, safe and reliable public transport. And it may be true that reversing the course of car-centred mobility could best be achieved in some measure by a return to many of the conditions of pre-car culture.

It feels nearly impossible to imagine what that might look like, but what we shouldn't do is feel paralysed by FOBO (fear of better options). An emphasis on the local, on smaller, walkable neighbourhoods and more central job and amenity centres is a good place to start. It's not just tiny villages that can reap the benefits of this spatial reconsideration; even the quintessential car city, Los Angeles, has been redefining itself as a city of neighbourhoods, investing millions in public transportation, including a 'subway to the sea'.

A subway to the sea sounds like something straight out of Birnin Zana, the city at the centre of the 2017 box-office hit *Black Panther*. Designers, wrote Brentin Mock, who covers social justice and equity for *City Lab*, 'have been wowed by [the nation of] Wakanda's mechanical marvels of hyperloop rapid transit, maglev trains, dragonfly-shaped spaceships, hoverbikes, skyscrapers orchestrated from chords of stone, wood, and metal, and other innovative spectacles'. In conjuring this place, he continues, its creators explored 'what the ideal model for equitable development looks like; how to preserve the traditions and culture of a place while embracing innovation and technology; how transit can co-mingle with walkability; and the role of design in facilitating spaces that protect vulnerable populations from oppressive forces'.[14] And, most notably, what a future without private transportation might look like. There are no cars in Wakanda [140].

Wakanda is the stuff of fiction – but it doesn't have to be. The success of *Black Panther* revealed a hunger for experiencing other narratives; cities need to tell new and different stories, and that will occur only when more people get to tell them.

So much public dialogue about cities, as the urbanism writer Alissa Walker has reminded us in 'Mansplaining the City',[15] 'is largely unchanged since [activist Jane] Jacobs's era, when her adversary Robert Moses dismissed the group of mostly female activists working to save [New York's] Washington Square Park as "a bunch of mothers"'.[16] Walker, who's been a force in the aforementioned shift towards a more walkable LA, observes that while those activists may finally be emerging from behind the scenes, to take on leadership positions, 'the shift towards just and equitable cities will only happen when a more diverse group of Americans are in positions to make policy decisions that shape our neighborhoods.'

This shift may prove most difficult in America, which is so inextricably linked to the car. Yet elsewhere around the world, car-free cities are becoming a real thing. Since 1974 Bogotá in Colombia has closed 75 miles of roads to vehicles one day every week, in an event known as *Ciclovía*; several other cities, including Los Angeles and San Francisco, now have their own version (albeit less frequently) [141]. Copenhagen in Denmark is building a bike superhighway; more than half of its residents cycle to work or school every day. Oslo in Norway is set to ban cars from its city centre by 2019; Madrid in Spain will do the same by 2020. Paris is banning diesel cars (as is London) and instituted car-free Sundays in 2016. Only 50 per cent of the roads in Chengdu in China

↑ **140** The fictional nation of Wakanda, home to Marvel's superhero Black Panther, imagines a city without private transportation

will allow vehicles, and the city was designed so that people can walk where they need to go within 15 minutes. Hamburg in Germany is creating a 'green network' of connected spaces that can be accessed without cars. Berlin has a 34-square-mile low-emission zone in its city centre, which bans all gas and diesel vehicles that fail to meet national emission standards. And the list goes on ...

For decades, the automobile provided a pathway to economic opportunity and upward mobility. But now the negative consequences – including a reliance on fossil fuels and increased emissions of greenhouse gases; a dramatic increase in the rate of deaths caused by cars;[17] the disconnection of local community and weakening of local economies; the rise in obesity and heart disease; congestion and sprawl; a lack of investment in non-car infrastructure; and increasing social segregation and isolation – seem to overwhelmingly outweigh the positives.

The challenge is not a technological one really, although ride-sharing apps, electric bikes and such all have their place. There are policies to change, zonings to reform and money to reallocate, but ultimately the real obstacle is a psychic one. We need to hear from a 'bunch of mothers' – and a bunch of other folks, too. We've imagined our futures behind the wheel for so long; the challenge now is to set out on our own two feet.

→ **141** In response to
congestion and pollution,
the city of Bogotá introduced
the event *ciclovías*. Each
Sunday and public holiday,
main streets are blocked
off to cars to make way for
pedestrians and cyclists. The
event has inspired *ciclovías*
in other cities such as LA,
depicted here.

THERE ARE NO CARS IN WAKANDA

NOTES

Introduction

1. Panofsky paraphrased by Hans Obrist, in Marta 2018, p. 10.

The Design of Speed

1. Mom 2004, p. 48.
2. Parissien 2013, p. 3.
3. Ibid.
4. Nicholson 1982.
5. Roberts 1998.
6. Hughes 1993.
7. Ibid.
8. Broyles 2013.
9. Loiperdinger 2004, pp. 89–118.
10. Gray 1991.
11. Mom 2004, p. 48.
12. Dougall 2013, p. 1.
13. 'How Brooklands Started', *Autocar* (17 August 1967), vol. 127, no. 3731, p. 43.
14. Gosling 2009.
15. Williams 2014, p. 110.
16. Chéroux 2017, p. 304.
17. Bachelor 1994, p. 19.
18. Wasef 2009, pp. 42–5.
19. Bröhl 1978.
20. Margolius and Henry 2015, p. 110.
21. Rieger 2013, p. 66.
22. Bel Geddes 1934, pp. 553–8.
23. Gartman 1994.
24. Ingrassia 2012, p. 159.
25. Acerbi 2012.
26. Hilton 2012.
27. Avery 2018, p. 6.
28. Marsh and Collett 1986, p. 155.
29. Ingrassia 2012, p. 129.
30. Ibid., p. 134.
31. Patton 2005, p. 29.
32. Ibid., p. 33.
33. Johnson 2009, p. xii.
34. National Center for Statistics and Analysis (NCSA), Motor Vehicle Traffic Crash Data Resource Page; https://crashstats.nhtsa.dot.gov/ (accessed 15 September 2018).
35. 'Reported Road Casualties Great Britain, Annual Report: 2017', Department for Transport (2017); https://www.gov.uk/government/statistics/reported-road-casualties-great-britain-annual-report-2017 (accessed 15 September 2018).

In My Ford Fiesta

1. Nader 1973, p. lxxiii.
2. *Freedom of the American Road*, directed by Leo Hurwitz (Mpo Productions Inc., 1956).
3. Kerouac 2000, p. 121.
4. Bel Geddes 1940, p. 10.
5. Levin 2000, p. 25.
6. College of Policing, 'National Police Driving Schools Conference: Police Driving 2019', National Police Library; www.library.college.police.uk/docs/NPDSC-Police-Driving-Manual-2009.pdf (accessed 17 December 2018).
7. 'Stephen Bayley on CRASHED CARS and the DEATH OF THE MODERN ERA'; *032c*; www.032c.com/stephen-bayley-crashed-cars-death-modern-era/ (accessed 17 December 2018).
8. Seiler 2008, p. 63.
9. Marinetti 1909.
10. *Exclusive: Charlie Sheen Says He's 'Not Bipolar but "Bi-Winning"'* (2011); www.youtube.com/watch?v=h5aSa4tmVNM (accessed 17 December 2018).

The Manufacture of Plenty

1. 'In Red Car to the Kingdom of Communism' is a New York Public Library translation of '*Na krasnom avtomobile revoliutsii v tsarstvo kommunizma*'.
2. Chapter title in Nevins and Hill 1954.
3. Wells 2007, p. 522.
4. Rieger 2013, p. 11.
5. Wells 2007, p. 497.
6. Figures from Motor magazine's 1907 Motor Car Directory. Published in Julian Chase, 'The Typical Motor-Car', *Cosmopolitan Magazine* (April 1907), vol. XLII, no. 6.
7. Purcell 1992, p. 156.
8. Ford Motor Company, *Factory Facts from Ford* (Detroit, 1915), p. 62.
9. 'Items of Interest', *Bush Advocate* (27 May 1910), vol. XXII, no. 121, p. 4.
10. Boris Shpotov has written about the problems that Ford encountered introducing their system of production in Soviet Russia, where they were trying to train workers who often had no factory experience. Shpotov 2008, p. 441.
11. Ford 1923, p. 108.
12. See Hounshell 1985, chapter 6, for a very detailed description of Ford's development of the assembly line, including production of the magneto flywheel seen image 43.
13. Ford 1923, p. 81. Moving conveyor belts had also been used in a number of other nineteenth-century factories, including the canning industry and for casting moulds at the Westinghouse foundry. For a detailed account of nineteenth-century precursors to Fordist production, see Hounshell 1985 and Giedion 1948.
14. In 1911 Taylor published a book, *The Principles of Scientific Management*, which outlined his theories of factory work and production.
15. See Grandin 2009 for an excellent history of the Fordlandia plantation.
16. Meyer (1926) 1994, p. 448.
17. Mazur 1931, p. 153.
18. David Hounshell states that 'Fordism' was used to mean highly mechanized production, a moving assembly line, high wages and low prices on products. Hounshell 1985, p. 11.
19. 'Obituary: Mr Thomas Bata', *Manchester Guardian* (13 July 1932), p. 4.
20. Letter from Dr Sun Yat-sen to Henry Ford, 12 June 1924. Henry Ford Museum: Acc.285:285:602.
21. As Sonia Melnikova-Raich explains, the Soviets first

entered into discussions with Ford about the local production of their tractors in 1923, planning to build a plant in Stalingrad. Before deciding on Ford, they also considered contracts with other American and German companies. Melnikova-Raich 2010, p. 58.

22. Hughes 1989, p. 278.
23. Stites 1989, pp. 148–9.
24. Siegelbaum 2008, p. 2.
25. Melnikova-Raich 2010, p. 60.
26. Mumford 1934, pp. 354–5.
27. Mumford pitted himself in opposition to 'frantic attempts that have been made in America by advertising agencies and "designers" to stylize machine-made objects'. Mumford 1934, p. 355.
28. Meyer wrote: 'To the semi-nomad of contemporary economic life the standardisation of residential, clothing, nutritional, and cultural requirements affords the vital quotient of mobility, economy, simplicity, and ease. *The degree of standardisation is the index of our collective economy'* (italics in the original text). Meyer (1926) 1994, p. 448. For a discussion of Meyer's cooperative projects, see Borra 2013.
29. Breuer (1928) 1994, p. 453.
30. Just two of many examples of architects using the image of the assembly line are Walter Gropius, whose plans for mass-produced houses were based on a study of Fordist production and were dubbed 'Wohnford' by Siegfried Giedion (Hughes 1989, p. 316); and Le Corbusier, who designed his 'Maison Citrohan' to be mass-produced using automotive-industry techniques and wrote: 'If houses were constructed by industrial mass production, like chassis, unexpected but

sane and defensible forms would soon appear, and a new aesthetic would be formulated with astonishing precision.' Le Corbusier 1925, p. 133.
31. Interview with Phil Stallings, spot welder at the Ford assembly plant on the South Side of Chicago. Published in Terkel 1974, p. 161.
32. Letter to Henry Ford from the wife of an assembly-line worker, 23 January 1914. Henry Ford Museum: 64.167.1.94.
33. Hounshell 1985, p. 11.
34. Raymond 2006, p. 45.
35. 'Unionism not Fordism' leaflet, box 8, folder 7, of the UAW President's Office: Walter P. Reuther Records. UAW archives, Walter P. Reuther Library, Wayne State University.
36. The best-known early use of this type of automated process in the car industry was the A.O. Smith Corporation's so-called 'Mechanical Marvel', a giant transfer machine put into operation in the 1920s, aiming to manufacture car frames without any human labour. Contemporary with the A.O. Smith machine, in 1923 the British engineer Frank G. Woolard developed a transfer machine used to manufacture engine blocks, gear-box castings and flywheels at Morris Engines in Coventry. For A.O. Smith, see Hounshell 2000 and Giedion 1948, pp. 118–20; for Morris, see Hounshell 2000, p. 106.
37. Hounshell 2000, pp. 101 and 116.
38. The patent was awarded in 1961: US Patent No. 2,988,237.
39. Flint 1970, p. 10.
40. Zaretsky 2007, pp. 105–6.
41. Gary Bryner, in Terkel 1974, p. 191.
42. Bryner described the Unimate as looking 'just like a praying mantis. It

goes from spot to spot to spot. It releases that thing and then it jumps back into place, ready for the next car.' Bryner, in Terkel 1974, p. 191.
43. Cusumano 1988.
44. Daito 2000.
45. Buck-Morss 2002, p. ix.

Almost Like a Car

1. Quintas 1994, pp. 25–47.
2. http://cristal.inria.fr/~weis/info/commandline.html (accessed 4 December 2018).
3. https://www.genivi.org/sites/default/files/BMW_Case_Study_Download_040914.pdf (accessed 4 December 2018).
4. https://www.stout.com/insights/report/2016-automotive-warranty-recall-report (accessed 4 December 2018).
5. https://electrek.co/2018/09/12/tesla-releasing-more-battery-capacity-free-supercharging-hurricane-florence/ (accessed 4 December 2018).
6. Star and Bowker 2002.
7. https://deadline.com/2016/02/oscars-mad-max-fury-road-colin-gibson-production-design-interview-1201695315/ (accessed 4 December 2018).
8. Darwin (1859) 2004.
9. https://www.dezeen.com/2015/05/23/mad-max-cornered-the-market-in-a-particular-vision-of-the-post-apocalyptic-future/ (accessed 4 December 2018).
10. Pinch and Bijker 1987, pp. 1–6.
11. Kline and Pinch 1996, pp. 763–95.
12. https://motherboard.vice.com/en_us/article/kz p7ny/tractor-hacking-right-to-repair (accessed 4 December 2018).
13. www.nytimes.com/2019/04/06/opinion/sunday/right-to-repair-elizabeth-warren-antitrust.html (accessed 12 April 2019).

Making the Modern Consumer

1. Flink 1976, pp. 142–3.
2. Ludvigsen 1995, pp. 51–9.
3. *Vogue Paris* (June 1931), p. 20a.
4. Sloan 1963, p. 72.
5. Slade 2007, pp. 29–56.
6. Blaszczyk 2012.
7. Ibid.
8. Ibid.
9. Sloan 1963, p. 265.
10. Pulos 1986.
11. *Motor* magazine (March 1927).
12. *Saturday Evening Post* (3 March 1927).
13. Armi 1988, p. 30.
14. General Motors Corporation, Cadillac Motor Car Company/LaSalle advertising, December 1927.
15. *Die Dame* (1922), vol. 24, p. 8.
16. Chrysler Corporation, advertising for Dodge *La Femme* (1955).
17. General Motors Corporation, press release (26 May 1957).
18. Berr 1904, p. 829.
19. *Wall Street Journal* (July 1959).
20. Nader 1965.
21. Lippincott in Packard 1960, p. 70.
22. Packard 1960.
23. Ant Farm 1976.
24. Miller 2001.
25. Runzheimer & Company, *Your Driving Costs* (Chicago 1960).
26. Whiteley 1987, pp. 3–27.
27. *Sales Management* (6 May 1960).

Oh Lord, Won't You Buy Me a Mercedes-Benz

1. Janis Joplin recorded 'Mercedes-Benz' on 1 October 1970.
2. See Grosse-Ophoff et al. 2017.
3. Little 2017.
4. Shapland 2018.
5. Transport for London, 'Roads Task Force – Technical Note 12. How Many Cars are There in London and Who Owns Them?' (2012); http://content.tfl.gov.uk/technical-note-12-how-many-cars-are-there-in-london.pdf

6. Shirouzu et al. 2018.
7. Clarke 1999, p. 68.
8. 'New Fiat 500 Collezione. Discover the Autumn/Winter Collection'; https://www.fiat.co.uk/fiat-500-range/500-collezione (accessed 28 October 2018).
9. Gao et al. 2016.
10. Shirouzu and Lienert 2018.
11. Adam Cohen in McMahon 2018.
12. https://patents.google.com/patent/US9272708B2/en (accessed 8 October 2018).
13. 'From Self-driving Cars to 3D Printing, BMW, Patricia Urquiola & UNStudio Predict the Future of Mobility', *Designboom* (25 July 2018); https://www.designboom.com/technology/bmw-patricia-urquiola-unstudio-future-mobility-07-25-2018/ (accessed 28 September 2018).
14. 'The Dwelling Lab by Patricia Urquiola, Giulio Ridolfo and BMW', press release from Kvadrat (April 2019); https://kvadrat.dk/collaborations/the-dwelling-lab-by-patricia-urquiola-giulio-ridolfo-and-bmw (accessed 28 September 2018).
15. 'Spaces on Wheels Report', Studio 10 (September 2018); https://space10.io/spaces-on-wheels-report/ (accessed 8 October 2018).
16. 'Rethinking the Urban House-share', mini.co.uk; https://www.mini.co.uk/en_GB/home/explore/mini-living-shanghai-press-release.html (accessed 8 October 2018).
17. 'Coming Soon: New Electric Cars', *What Car?* (26 January 2018); https://www.whatcar.com/news/coming-soon-new-electric-cars/n17030 (accessed 8 October 2018).
18. *Rams*, directed by Gary Hustwit, 2018.

The Race to Extraction

1. '*Dans l'après-midi du 12 décembre 1955, à la suite d'un petit accident dont la cause est resté inconnue, une tempête électrique, une tournade … se déchaina sur tout l'Ouest de l'Europe et amena … des profondes perturbations à la vie générale.*' Robida 1892, pp. 1–2.
2. Dutch feminist Aletta Jacobs, commenting on cars seen during a visit to Paris in 1890. Quoted in Mom 2015, p. 96.
3. Ibid., p. 62.
4. Lavergne 1900, pp. 259–60.
5. 'Columbia Motor Carriages', *Scientific American* (13 May 1899), p. 304.
6. See Scharff 1992, chapter 3.
7. General Electric advertisement, quoted in ibid., p. 39.
8. Quoted in Mom 2004, p. 44.
9. See Mom 2004 and 2015.
10. Levitt 1909, p. 32.
11. King Leopold was a famous automobile enthusiast, with the extravagance of his cars eagerly watched and reported on. This obsession had huge and horrific consequences in Leopold's colonial ambitions. The Belgian king was responsible for invading the Congo Free State in the 1890s, colonizing the country by illegally taking over its land for the cultivation of rubber. Leopold hoped to capitalize on a global market that was growing exponentially, primarily through the car-tyre industry. Huge numbers of Congolese were enslaved to work on Belgian rubber plantations, with the atrocious conditions resulting in an effective genocide.
12. See, for example, the 'Automobile News' column in the following issues of *Scientific American*: 21 April 1900, p. 250 (Sudan); 6 April 1901, p. 214 (Algeria); 18 May 1901, p. 310 (Madagascar); 14 December 1901, p. 392 (Congo Free State).
13. Despatches from the Royal East Africa Automobile Association, 1924. National Archives: CO 533/309.
14. Levine 2005, p. 83.
15. Ibid., p. 81.
16. Clarsen 2015, pp. 165–86.
17. Radó 1938, p. 104.
18. Smil 2008, p. 3.
19. 'The Alcohol Motor and its Possibilities', *Scientific American* (11 January 1902), p. 18.
20. Quoted in Winegard 2016, p. 92.
21. See Scazzieri 2015, pp. 25–45.
22. The Mexican oil industry was nationalized in the 1930s, limiting multi-national investment. Isser 1996, p. 9.
23. Shell, for example, released a series of advertisements in the early 1930s showing famous British landmarks. The series was entitled 'See Britain First on Shell'.
24. Radó 1938, p. 104.
25. Advertising tagline for the Esso Oil company, 1950.
26. Huber 2013, p. 73.
27. Smil 2017, p. 276.
28. Ibid., p. 278, and Smil 2008, p. 28.
29. Flink 1988, p. 287.
30. Huber 2013, pp. 74–6.
31. Ibid., p. 72.
32. From a television special celebrating the 75th anniversary of Standard Oil, quoted in Huber 2013, p. 72.
33. Tagline for an advertisement for Humble Oil, 1962.
34. Flink 1988, pp. 386–8.
35. SUVs made up more than 50 per cent of the US market in the 1990s. Smil 2008, p. 8; UN statistics: http://www.ipcc.ch/ipccreports/tar/wg3/index.php?idp=99 (accessed 10 November 2018).
36. Intergovernmental Panel on Climate Change, *Summary for Policymakers, in Global Warming of 1.5°C. An IPCC Special Report on the Impacts of Global Warming of 1.5°C Above Pre-industrial Levels and Related Global Greenhouse Gas Emission Pathways, in the Context of Strengthening the Global Response to the Threat of Climate Change, Sustainable Development, and Efforts to Eradicate Poverty,* ed. v. Masson-Delmotte et al., World Meteorological Organization (Geneva, Switzerland, 2018).
37. According to the UN, around 20 per cent of global emissions are produced by the transportation sector. Statistics: http://www.ipcc.ch/ipccreports/tar/wg3/index.php?idp=99 (accessed 10 November 2018).
38. The total number of electric cars is still a tiny percentage of overall car production: about 3 million out of 1.1 billion, in 2018. But with markets like China (currently the world's largest car manufacturer) looking to actively grow the production of battery-powered cars, this looks likely to increase significantly. Butler 2018.

Water Wars and the Miracle Metal

1. 'Orocobre: Operations. Salar de Olaroz'; https://www.orocobre.com/operations/salar-de-olaroz/ (accessed 17 October 2018).
2. BloombergNEF Electric Vehicle Outlook (2018); https://about.bnef.com/electric-vehicle-outlook/ (accessed 18 October 2018).
3. Centenera 2017.
4. Iñurrieta 2017.
5. 'Bolivia quiere ser la Arabia Saudí del litio y avisa: "Vamos a poner el precio a todo el mundo"', *Efe/El Economista* (January 2018); https://www.eleconomista.es/materias-primas/noticias/8856549/01/18/Bolivia-quiere-ser-la-Arabia-Saudi-del-Litio-y-avisa-Vamos-a-poner-el-precio-para-a-todo-el-

mundo.html; https://www.
energiminas.com/evo-
morales-bolivia-pondra-
el-precio-de-baterias-de-
litio-en-todo-el-mundo/
(accessed 16 October 2018).

6. Castilla 2017.
7. Clayton 2018.
8. 'Morales to End 500 Years
of Injustice', Al Jazeera
(January 2006); https://
www.aljazeera.com/
archive/2006/01/
200841015924784754.html
(accessed 16 October 2018).
9. Sherwood 2018a.
10. Sherwood 2018b.
11. Jamasmie 2018.
12. Els 2018.
13. Davenport et al. 2018.
14. Shabalala 2018.

Driving the Nation

1. Bennett and Vranica 2012.
2. Kroplick and Velocci 2008.
3. Moraglio 2017, p. 46.
4. Ibid., p. 53.
5. Ibid., p. 66.
6. Ibid., p. 139.
7. Rieger 2013, pp. 49–50.
8. Moraglio 2017, p. 160.
9. Baldwin et al. 2002.
10. DiMento and Ellis 2013.
11. Wilkins 2011.
12. Bachelor 1994, pp. 77–8.
13. Ibid., p. 76.
14. Rieger 2013, p. 42.
15. Ibid., p. 58.
16. Ibid., p. 59.
17. Ibid., pp. 71–2.
18. Ibid., p. 81.
19. Lottman 2003.
20. Michelin, propriétaires-
éditeurs, *The Michelin
Guide: More than 100 Years
of Experience* (Clermont-
Ferrand 2007).
21. Ibid., p. 90.
22. Ibid., p. 82.
23. Colafranceschi 2013, p. 61.
24. Ibid., p. 62.
25. Parissien 2013, p. 189.
26. The ad agency DDB started
producing promotional
material for Volkswagen in
1959, notoriously flaunting
convention by positioning
the car outside the norms
of car advertising, using

humour and irony to create
a unique identity; see Rieger
2013, pp. 212–14.
27. Bullen 2015, p. 137.

Political Symbolism and
the Nyayo Pioneer Car

1. 'Cheers as President
Launches Nyayo Pioneer
Car' (1990), *Daily Nation*
(accessed 1 October 2018).
2. Hempstone 1997.
3. 'Cheers as President
Launches Nyayo Pioneer
Car' (1990), *Daily Nation*
(accessed 1 October 2018).

There Are No Cars
in Wakanda

1. http://unevenearth.
org/2018/08/the-social-
ideology-of-the-motorcar/
(accessed 25 August 2018).
2. https://www.theatlantic.
com/ideas/archive/2018/09/
how-self-driving-cars-
could-ruin-the-american-
city/569518/ (accessed
25 September 2018).
3. https://www.topspeed.
com/cars/others/2018-
kalashnikov-cv-1-russian-
ev-ar182341.html (accessed
31 August 2018).
4. http://www.findpatent.ru/
patent/242/2428364.html
(accessed 1 September 2018).
5. https://www.treehugger.
com/urban-design/
future-you-may-never-get-
out-your-car-shop.html
(accessed 31 August 2018).
6. See SPACE10, a Future Living
Lab; https://space10.io
(accessed 31 August 2018).
7. Volvo describes its
autonomous vehicle concept
as 'what becomes possible
when we remove the human
driver, using new freedoms
in design and recapturing
time'; https://www.media.
volvocars.com/us/en-us/
media/pressreleases/237020/
volvo-cars-new-360c-
autonomous-concept-
reimagining-the-work-
life-balance-and-the-
future-of-cities (accessed

31 August 2018).
8. Elon Musk, Tesla at the
Neural Information
Processing Systems
Conference in 2017; https://
www.wired.com/story/
elon-musk-awkward-dislike-
mass-transit/ (accessed
25 September 2018).
9. Tori Venable of Americans
for Prosperity, quoted in
Tabuchi 2018.
10. The 'sixth extinction' refers
to the current worldwide
loss of biodiversity: the sixth
time in world history that
a large number of species
have disappeared in
unusually rapid succession.
This extinction is caused
not by asteroids or ice ages
but by humans.
11. https://www.theguardian.
com/world/2017/jun/15/
timothy-morton-
anthropocene-philosopher
(accessed 25 September 2018).
12. https://www.theverge.
com/2018/7/19/17586878/
larry-page-flying-car-
opener-kitty-hawk-cora
(accessed 25 September 2018).
13. See Gerding Edlen's
Principles of Place; https://
www.gerdingedlen.com/
principles-of-place/make-
20-minute-living-real/
(accessed 2 December 2018).
14. https://www.citylab.
com/equity/2018/02/the-
wakanda-reader/553865/
(accessed 15 September 2018).
15. https://www.curbed.
com/2017/8/16/16151000/
mansplain-gentrification-
define-richard-florida-
saskia-sassen (accessed
30 September 2018).
16. https://www.theguardian.
com/books/2009/sep/12/
jane-jacobs-new-york-
history (accessed
25 September 2018).
17. https://www.nsc.org/
road-safety/safety-topics/
fatality-estimates (accessed
9 May 2018).

BIBLIOGRAPHY

Acerbi 2012
Leonardo Acerbi, *Mille Miglia Story 1927–1957* (Vimodrone, Milan 2012)

Ant Farm 1976
Ant Farm, *Automerica: A Trip Down US Highways from World War II to the Future* (New York 1976)

Armi 1988
C. Edson Armi, *The Art of American Car Design: The Profession and Personalities* (Pennsylvania 1988)

Avery 2018
Matt Avery, *COPO Camaro, Chevelle & Nova: Chevrolet's Ultimate Muscle Cars* (Forest Lake, MN 2018)

Bachelor 1994
Ray Bachelor, *Henry Ford: Mass Production, Modernism, and Design* (Manchester 1994)

Baldwin et al. 2002
Peter Baldwin, Robert Baldwin and Dewi Ieuan Evans, *The Motorway Achievement* (London 2002)

Bel Geddes 1934
Norman Bel Geddes, 'Streamlining', *Atlantic Monthly* (November 1934), no. 154

Bel Geddes 1940
Norman Bel Gedes, *Magic Motorways* (New York 1940)

Bennett and Vranica 2012
Jeff Bennett and Suzanne Vranica, 'Chrysler Dealers Defend "Halftime in America" Ad', *Wall Street Journal* (9 February 2012); https://www.wsj.com/articles/SB10001424052970204136045772113917192371 60 (accessed 9 November 2018)

Berr 1904
Emile Berr, 'Une Exposition Parisienne – Le "Salon" des Chauffeurs', *Revue Bleu* (24 December 1904), vol. 11

Blaszczyk 2012
Regina Lee Blaszczyk, *The Color Revolution* (London 2012)

Borra 2013
Bernardina Borra, 'Hannes Meyer: Co-op Architecture', *San Rocco 6 / Collaborations* (2013)

Breuer (1928) 1994
Marcel Breuer, 'Metal Furniture and Modern Spatiality' (1928),

in Anton Kaes, Martin Jay and Edward Dimendberg (eds), *The Weimar Republic Sourcebook* (Berkeley, Los Angeles and London 1994)

Bröhl 1978
Hans-Peter Bröhl, *Paul Jaray, Stromlinienpionier: von der Kastenform zur Stromlinienform* (Bern 1978)

Broyles 2013
Susannah Broyles, 'Vanderbilt Ball: How a Costume Ball Changed New York Elite Society', *MCNY Blog: New York Stories* (6 August 2013); https://blog.mcny.org/2013/08/06/vanderbilt-ball-how-a-costume-ball-changed-new-york-elite-society/ (accessed 18 September 2018)

Buck-Morss 2002
Susan Buck-Morss, *Dreamworld and Catastrophe: The Passing of Mass Utopia in East and West* (Cambridge, MA and London 2002)

Bullen 2015
George Bullen, *Nine Months in Iran* (Morrisville, NC 2015)

Butler 2018
Nick Butler, 'Why the Future of the Electric Car Lies in China', *Financial Times* (17 September 2018); https://www.ft.com/content/1c31817e-b5a4-11e8-b3ef-799c8613f4a1 (accessed 30 November 2018)

Castilla 2017
Juliana Castilla, 'Argentina Seeks to Overtake Chile in South America Lithium Race', *Reuters* (8 November 2017); https://uk.reuters.com/article/uk-argentina-mining-lithium/argentina-seeks-to-overtake-chile-in-south-america-lithium-race-idUKKBN1DD2IP (accessed 18 October 2018)

Centenera 2017
Mar Centenera, 'Las inundaciones golpean el centro de Argentina y amenazan la cosecha de soja', *El País* (18 January 2017); https://elpais.com/internacional/2017/01/17/argentina/1484690353_751154.html (accessed 18 October 2018)

Chéroux 2017
Clément Chéroux, 'In Praise of the Photographic Accident', *Auto Photo*, Fondation Cartier pour l'art contemporain (Paris 2017)

Clarke 1999
Sally Clarke, 'Managing Design: The Art and Colour Section at General Motors, 1927–1941', *Journal of Design History* (1999), vol. 12, no. 1

Clarsen 2015
Georgine W. Clarsen, 'Mobile Encounters: Bicycles, Cars, and Australian Settler Colonialism', *History Australia* (2015), vol. 12, no. 1

Clayton 2018
Frederick Clayton, 'As Others Snub Bolivia's Lithium, Will Morales' Gamble on Germany Pay Off?', *Americas Quarterly Online* (May 2018); https://www.americasquarterly.org/content/others-snub-bolivias-lithium-will-morales-gamble-germany-pay (accessed 16 October 2018)

Colafranceschi 2013
Simone Colafranceschi, 'The Ebb and Flow of the Autogrill', in Pippo Ciorra (ed.), *Energy: Oil and Post-Oil Architecture and Grids* (Milan 2013)

Le Corbusier 1925
Le Corbusier, *Towards a New Architecture* (Paris 1925)

Cusumano 1988
Michael A. Cusumano, 'Manufacturing Innovation: Lessons from the Japanese Auto Industry', *MIT Sloan Management Review* (15 October 1988)

Daito 2000
Eisuke Daito, 'Automation and the Organisation of Production in the Japanese Automobile Industry: Nissan and Toyota in the 1950s', *Enterprise and Society* (March 2000), vol. 1, no. 1, pp. 139–78

Darwin (1859) 2004
Charles Darwin, *On the Origin of Species* (1859), (London 2004)

Davenport et al. 2018
Emily Davenport, Christine Folch and Connor Vasu, *Itaipú Dam: Paraguay's Growth Potential* (2018);

https://itaipupost2023.files.
wordpress.com/2018/06/
white-paper-final-draft-itaipu-
paraguays-growth-potential2.
pdf (accessed 16 October 2018)

DiMento and Ellis 2013
Joseph F. DiMento and Cliff
Ellis, *Changing Lanes: Visions
and Histories of Urban Freeways*
(Cambridge, MA 2013)

Dougall 2013
Angus Dougall, *The Greatest
Racing Driver* (Bloomington,
IN 2013)

Els 2018
Frik Els, 'Lithium Price:
Chile Giant's Scorched Earth
Strategy', *Mining.com* (17 August
2018); http://www.mining.com/
lithium-price-chile-giants-
scorched-earth-strategy/
(accessed 17 October 2018)

Flink 1976
James J. Flink, *The Car Culture*
(London 1976)

Flink 1988
James J. Flink, *The Automobile
Age* (Cambridge, MA and
London 1988)

Flint 1970
Jerry M. Flint, 'Auto Industry
Struggling to Stop Lag in
Productivity', *New York Times*
(8 August 1970)

Ford 1923
Henry Ford, in collaboration
with Samuel Crowther, *My Life
and Work* (New York 1923)

Gao et al. 2016
Paul Gao, Hans-Werner Kaas,
Detlev Mohr and Dominik Wee,
'Disruptive Trends that Will
Transform the Auto Industry',
McKinsey & Company (January
2016); https://www.mckinsey.
com/industries/automotive-
and-assembly/our-insights/
disruptive-trends-that-will-
transform-the-auto-industry
(accessed 28 September 2018)

Gartman 1994
David Gartman, *Auto Opium:
A Social History of American
Automobile Design* (New
York 1994)

Giedion 1948
Siegfried Giedion,
*Mechanization Takes Command:
A Contribution to Anonymous

History* (New York 1948)

Gosling 2009
Peter Gosling, *Quest For Speed –
Simple Guides* (London 2009)

Grandin 2009
Greg Grandin, *Fordlandia: The
Rise and Fall of Henry Ford's
Forgotten Jungle City* (New
York 2009)

Gray 1991
Edwyn Gray, *The Devil's Device:
Robert Whitehead and the History
of the Torpedo* (Annapolis,
MD 1991)

Grosse-Ophoff et al. 2017
Anne Grosse-Ophoff, Saskia
Hausler, Kersten Heineke and
Timo Möller, 'How Shared
Mobility Will Change the
Automotive Industry', *McKinsey
& Company* (April 2017); https://
www.mckinsey.com/industries/
automotive-and-assembly/
our-insights/how-shared-
mobility-will-change-the-
automotive-industry (accessed
28 September 2018)

Hempstone 1997
Smith Hempstone, *Rogue
Ambassador: An African Memoir*
(Sewanee, TN 1997)

Hilton 2012
Christopher Hilton, *Le Mans '55:
The Crash That Changed the
Face of Motor Racing*
(Nottingham 2012)

Hounshell 1985
David Hounshell, *From the
American System to Mass
Production 1800–1932* (Baltimore,
MD and London 1985)

Hounshell 2000
David Hounshell, 'Automation,
Transfer Machinery, and
Mass Production in the US
Automobile Industry in
the Post-World War II Era',
Enterprise & Society (March
2000), vol. 1, no. 1, pp. 100–38

Huber 2013
Matthew T. Huber, *Lifeblood: Oil,
Freedom and the Forces of Capital*
(Minneapolis, MN 2013)

Hughes 1989
Thomas P. Hughes, *American
Genesis: A Century of Invention
and Technological Enthusiasm,
1870–1970* (Chicago 1989)

Hughes 1993
Thomas Park Hughes, *Networks
of Power: Electrification in
Western Society, 1880–1930*
(Baltimore, MD 1993)

Ingrassia 2012
Paul Ingrassia, *Engines of
Change* (New York 2012)

Iñurrieta 2017
Sebastián Iñurrieta, 'Macri:
"[Vamos a lanzar el] Plan
Belgrano, que arranca de
US$16 mil millones de
inversión en infraestructura"',
Chequeado (10 December
2017); http://chequeado.
com/ultimas-noticias/
macri-vamos-a-lanzar-el-plan-
belgrano-que-arranca-de-us16-
mil-millones-de-inversion-en-
infraestructura-2017/ (accessed
18 October 2018)

Isser 1996
Steve Isser, *The Economics and
Politics of the United States Oil
Industry, 1920–1990* (London
and New York 1996)

Jamasmie 2018
Cecilia Jamasmie, 'Australia
Takes Over Chile as World's No.1
Lithium Producer', *Mining.com*
(June 2018); http://www.mining.
com/australia-takes-chile-
worlds-no-1-lithium-producer/
(accessed 17 October 2018)

Johnson 2009
Ann Johnson, *Hitting the
Brakes: Engineering Design and
the Production of Knowledge*
(Durham, NC 2009)

Kerouac 2000
Jack Kerouac, *On The Road*
(London 2000)

Kline and Pinch 1996
Ronald Kline and Trevor
Pinch, 'Users as Agents of
Technological Change: The
Social Construction of the
Automobile in the Rural United
States', *Technology and Culture*
(1996), vol. 37, no. 4

Kroplick and Velocci 2008
Howard Kroplick and Al Velocci,
The Long Island Motor Parkway
(Mount Pleasant, SC 2008)

Lavergne 1900
Gérard Lavergne, *Manuel
théorique et pratique de
l'automobile sur route* (Paris 1900)

Levin 2000
Miriam R. Levin, 'Contexts of
Control', in Miriam R. Levin
(ed.), *Cultures of Control*
(Amsterdam 2000)

Levine 2005
Alison Murray Levine, 'Film and
Colonial Memory: *La Croisière
noire* 1924–2004', in Alec G.
Hargreaves (ed.), *Memory,
Empire, and Postcolonialism:
Legacies of French Colonialism*
(Oxford 2005)

Levitt 1909
Dorothy Levitt, *The Woman and
the Car: A Chatty Little Handbook
for All Women Who Motor or Who
Want to Motor* (London 1909)

Little 2017
Stephen Little, 'Uber Data
Leak Hit 2.7m UK Customers,
Admits Ride-Hailing
Company', *The Independent*
(29 November 2017); https://
www.independent.co.uk/news/
business/news/uber-data-
leak-customers-ride-hailing-
company-taxi-app-london-
ban-a8082566.html (accessed
8 October 2018)

Loiperdinger 2004
Martin Loiperdinger, 'Lumiere's
Arrival of the Train: Cinema's
Founding Myth', *The Moving
Image* (Spring 2004), vol. 4,
no. 1, pp. 89–118

Lottman 2003
Herbert Lottman, *The Michelin
Men: Driving an Empire*
(London 2003)

Ludvigsen 1995
Karl Ludvigsen, 'A Century of
Automobile Body Evolution',
Automotive Engineering (1995),
vol. 103

McMahon 2018
Jeff McMahon, 'How Car-
sharing Companies Can
Save the World from the
Autonomous Vehicle Robot
Apocalypse', *Forbes* (18 March
2018); https://www.forbes.com/
sites/jeffmcmahon/2018/03/18/
how-car-sharing-companies-
can-save-the-world-from-the-
autonomous-vehicle-robot-
apocalypse/#217ca9e4fab7
(accessed 28 September 2018)

Margolius and Henry 2015
Ivan Margolius and John G.
Henry, *Tatra – The Legacy of Hans
Ledwinka* (Poundbury 2015)

Marinetti 1909
F.T. Marinetti, 'Manifeste
du futurisme', *Le Figaro*
(20 February 1909)

Marsh and Collett 1986
Peter Marsh and Peter Collett,
*Driving Passion: The Psychology
of the Car* (London 1986)

Marta 2018
Karen Marta (ed.), *Hans Ulrich
Obrist: The Athens Dialogues*
(London 2018)

Mazur 1931
Paul Mazur, 'The Doctrine
of Mass Production Faces a
Challenge', *New York Times*
(29 November 1931)

Melnikova-Raich 2010
Sonia Melnikova-Raich,
'The Soviet Problem with
Two "Unknowns": How an
American Architect and a Soviet
Negotiator Jump-started the
Industrialization of Russia,
Part I: Albert Kahn', *IA. The
Journal of the Society for
Industrial Archaeology* (2010),
vol. 36, no. 2

Meyer (1926) 1994
Hannes Meyer, 'The New World'
(1926), in Anton Kaes, Martin
Jay and Edward Dimendberg
(eds), *The Weimar Republic
Sourcebook* (Berkeley, Los
Angeles and London 1994)

Miller 2001
Daniel Miller, *Car Cultures*
(New York 2001)

Mom 2004
Gijs Mom, *The Electric Vehicle:
Technology and Expectations in
the Automobile Age* (Baltimore,
MD 2004)

Mom 2015
Gijs Mom, *Atlantic
Automobilism: Emergence and
Persistence of the Car*, 1895–1940
(New York and Oxford 2015)

Moraglio 2017
Massimo Moraglio, *Driving
Modernity: Technology, Experts,
Politics and Fascist Motorways,
1922–1943* (New York 2017)

Mumford 1934
Lewis Mumford, *Technics and
Civilization* (New York 1934)

Nader (1965) 1973
Ralph Nader, *Unsafe at Any
Speed: The Designed-in Dangers
of the American Automobile*
(Chicago 1965; reprinted 1973)

Nevins and Hill 1954
Allan Nevins and Frank Ernest
Hill, *Ford: The Times, the Man,
the Company* (New York 1954)

Nicholson 1982
T.R. Nicholson, 'Prelude to
Battle, 1895–6', in *The Birth of
the British Motor Car 1769–1897*
(London 1982)

Packard 1960
Vance Packard, *The Waste
Makers* (Philadelphia 1960)

Parissien 2013
Steven Parissien, *The Life of the
Automobile: A New History of the
Motor Car* (London 2013)

Patton 2005
Phil Patton, 'Design for
Destruction', in Paola Antonelli
(ed.), *Safe: Design Takes on Risk*
(New York 2005)

Pinch and Bijker 1987
Trevor J. Pinch and Wiebe E.
Bijker, 'The Social Construction
of Facts and Artifacts',
*The Social Construction of
Technological Systems: New
Directions in the Sociology and
History of Technology* (1987),
vol. 17

Pulos 1986
Arthur J. Pulos, *American
Design Ethic: A History of
Industrial Design to 1940*
(Massachusetts 1986)

Purcell 1992
Edward A. Purcell, Jr, *Litigation
and Inequality* (New York and
Oxford 1992)

Quintas 1994
Paul Quintas, 'Programmed
Innovation? Trajectories
of Change in Software
Development', *Information
Technology & People* (1994),
vol. 7, issue 1

Radó 1938
Alexander Radó, *The Atlas
of To-day and To-morrow*
(London 1938)

Raymond 2006
Francis L. Raymond, *This Day in
Business History* (New York 2006)

Rieger 2013
Bernhard Rieger, *The People's
Car: A Global History of the
Volkswagen Beetle* (Cambridge,
MA and London 2013)

Roberts 1998
Ian Roberts, 'Reducing Road
Traffic Would Improve Quality
of Life as well as Preventing
Injury', *British Medical Journal*
(1998)

Robida 1892
Albert Robida, *La vie électrique:
le vingtième siècle* (Paris 1892)

Scazzieri 2015
Luigi Scazzieri, 'Britain,
France, and Mesopotamian
Oil, 1916–1920', *Diplomacy &
Statecraft* (2015), vol. 26, no. 1

Scharff 1992
Virginia Scharff, *Taking the
Wheel: Women and the Coming
of the Motor Age* (Albuquerque,
NM 1992)

Seiler 2008
Cotten Seiler, *Republic of
Drivers: A Cultural History
of Automobility in America*
(Chicago 2008)

Shabalala 2018
Zandi Shabalala, 'Solid Demand
to Underpin Lithium as Price
Slides in 2018', *Reuters*
(17 September 2018); https://
www.reuters.com/article/
us-lithium-chemicals-prices/
solid-demand-to-underpin-
lithium-as-price-slides-in-
2018-idUSKCN1LX1PF
(accessed 17 October 2018)

Shapland 2018
Mark Shapland, 'Zipcar Expands
its New One-way Drive-
and-drop Service', *Evening
Standard* (15 January 2018);
https://www.standard.co.uk/
business/zipcar-expands-its-
new-oneway-driveanddrop-
service-a3740546.html
(accessed 08 October 2018)

Sherwood 2018a
Dave Sherwood, 'In Chilean
Desert, Global Thirst for
Lithium is Fueling a "Water
War"', *Reuters* (29 August 2018);
https://www.reuters.com/
article/us-chile-lithium-water/
in-chilean-desert-global-thirst-
for-lithium-is-fueling-a-water-
war-idUSKCN1LE16T (accessed
17 October 2018)

Sherwood 2018b
Dave Sherwood, 'Lithium
Miners' Dispute Reveals
Water Worries in Chile's
Atacama Desert', *Reuters*
(18 October 2018); https://
graphics.reuters.com/CHILE-
LITHIUM/010080VB1MH/index.
html (accessed 18 October 2018)

Shirouzu and Lienert 2018
Norihiko Shirouzu and Paul
Lienert, 'China Ride-hailing
Giant Didi Eyes Purpose-built
Fleet as Auto Market Shifts',
Reuters (24 April 2018); https://
uk.reuters.com/article/
uk-autoshow-beijing-didi/
china-ride-hailing-giant-
didi-eyes-purpose-built-
fleet-as-auto-market-shifts-
idUKKBN1HV061 (accessed
28 September 2018)

Shirouzu et al. 2018
Norihiko Shirouzu, Adam
Jourdan and Naomi Tajitsu,
'China's Didi Sets Up Electric
Car-sharing Platform', *Reuters*
(7 February 2018); https://
uk.reuters.com/article/
uk-renault-china-didi/
chinas-didi-sets-up-electric-
car-sharing-platform-
idUKKBN1FR0MI (accessed
8 October 2018)

Shpotov 2008
Boris M. Shpotov, 'The Case of
US Companies in Russia-USSR:
Ford in 1920s–1930s', in Hubert
Bonin et al., *American Firms in
Europe (1880–1980)* (Paris 2008)

Siegelbaum 2008
Lewis H. Siegelbaum, *Cars for
Comrades: The Life of the Soviet
Automobile* (Ithaca, NY and
London 2008)

Slade 2007
Giles Slade, *Made to Break:
Technology and Obsolescence in
America* (Massachusetts 2007)

Sloan 1963
Alfred Sloan, *My Years With
General Motors* (Massachusetts
1963)

Smil 2008
Vaclav Smil, *Oil: A Beginner's
Guide* (Oxford 2008)

Smil 2017
Vaclav Smil, *Energy and Civilisation: A History* (Cambridge, MA and London 2017)

Star and Bowker 2002
Susan Leigh Star and Geoffrey C. Bowker, 'How to Infrastructure', in Leah A. Lievrouw and Sonia Livingstone (eds), *The Handbook of New Media: Social Shaping and Consequences of ICTs* (London 2002)

Stites 1989
Richard Stites, *Revolutionary Dreams: Utopian Vision and Experimental Life in the Russian Revolution* (New York 1989)

Tabuchi 2018
Hiroku Tabuchi, 'How the Koch Brothers Are Killing Public Transit Projects Around the Country', *New York Times* (19 June 2018); https://www.nytimes.com/2018/06/19/climate/koch-brothers-public-transit.htm (accessed 1 July 2018)

Terkel 1974
Studs Terkel, *Working: People Talk About What They Do All Day and How They Feel About What They Do* (New York 1974)

Wasef 2009
Basem Wasef, *Legendary Race Cars* (Minneapolis, MN 2009)

Wells 2007
Christopher W. Wells, 'The Road to the Model T: Culture, Road Conditions, and Innovation at the Dawn of the American Motor Age', *Technology and Culture* (July 2007), vol. 48, no. 3

Whiteley 1987
Nigel Whiteley, 'Towards a Throw Away Culture', *Oxford Art Journal* (1987), vol. 10, no. 2

Wilkins 2011
Mira Wilkins, *American Business Abroad: Ford on Six Continents* (Cambridge 2011)

Williams 2014
Jean Williams, *A Contemporary History of Women's Sport, Part One* (New York 2014)

Winegard 2016
Timothy C. Winegard, *The First World Oil War* (Toronto 2016)

Zaretsky 2007
Natasha Zaretsky, *No Direction Home: The American Family and the Fear of National Decline, 1968–1980* (Durham, NC 2007)

ACKNOWLEDGEMENTS

Exhibitions are incredibly complex projects and one of the finest examples of collaboration on a large scale. For that reason, we would like to take the time here to extend our sincere gratitude to the many people who have made this exhibition and catalogue possible in one way or another.

First, thank you to Bosch for its generous support of the exhibition and its key loan of early Bosch safety equipment. Its early enthusiasm for the overall exhibition narrative that we have endeavoured to tell was deeply encouraging.

We would like to thank the lenders for making the exhibition rich and diverse in objects of all kinds: Associazione THAYAHT e RAM; Autostadt GMBH; Beaulieu – National Motor Museum; Robert Bosch GMBH; Bread & Salt Archive; British Library; Brooklands Museum; Burberry Limited; FCA Heritage – Fiat Chrysler Automobiles; Collection SEEBER MICHAHELLES; Cooper Hewitt – Smithsonian Design Museum; Countway Library of Medicine at Harvard; the Cranbrook Art Museum; Cranbrook Center for Collections and Research; Daimler Art Collection; Dallas Museum of Art; Kunstgewerbemuseum; Museum at the Fashion Institute of Technology; Fiat Chrysler Automobiles; Ford Heritage Fleet; Fondation Le Corbusier; François Schuiten; General Motors Archive; General Motors; Hagley Museum; Harry Ransom Humanities Research Center at the University of Texas; The Henry Ford; Honolulu Museum of Art; Institute of Mechanical Engineers; Italdesign; Le Musée National de la Voiture; Library of Congress; Louwman Museum – The Hague (NL);

MAXXI – the National Museum of the Arts of XXI Century – Rome; Metropolitan Museum of Art; Michelin Archives; Musée du Quai Branly; Museum of the City of New York; Moscow Design Museum; Mullin Automotive Museum; National Archives – DC; National Archives – Kew; National Portrait Gallery; National Technical Museum of the Czech Republic; Ne boltai! Collection; Schusev State Museum of Architecture; Science Museum Group; Smithsonian Institute: American Art Museum; Smithsonian Institute: National Museum of American History; Syd Mead – www.oblagon@sydmead.com; Targol Farazandeh, Tarlan Rafiee; Transport Accident Commission and Patricia Piccinini; the Valadez Family; Volvo Museum; and Walter P. Reuther Library, the Archives of Labor and Urban Affairs at Wayne State University.

For permission to reproduce their work in the exhibition, we would like to thank: Patrice Garcia, Olalekan Jeyifous x Ikiré Jones, Larry Gormley, the Estate of Raymond Loewy and Volkswagen Aktiengesellschaft.

Over coffees and lunches, in car rides and in meeting rooms, many people have advised and given feedback on our exhibition and we have cherished their input. This includes Jason Barlow, Christo Dantini, Marc Greuther, Kieran Hedigan, Marco Iezzi, Kold Lus, Craig Metros, Lord Montagu of Beaulieu, Natalie Morath, Claudia Porta, Fred Scharmen, Chris Thorpe, Thomas Turner and Liam Young.

Thanks to the contributing writers for this book, who approached our brief with great enthusiasm and originality, and who have made the content of the book all the richer: Johanna

Agerman Ross, Allison Arieff, Laurence Blair, Nanjala Nyabola, Oli Stratford and Georgina Voss.

This project has sat across two museum departments. First the Furniture, Textiles and Fashion Department (FTF), and later the Design, Architecture and Digital Department (DAD), and we are grateful to both departments for their support. In particular, a thank you to FTF Keeper Christopher Wilk for some early sage advice, and to DAD Keeper Christopher Turner for his continued support as the project moved across the finish line. And a special thank you to Senior Curator Corinna Gardner, who was an advisor and advocate for the project long before we began work on it.

Other colleagues across the museum have been a great help in sourcing V&A objects for the show, including: Julius Bryant, Zorian Clayton, Richard Edgcumbe, Ruth Hibbard, Kirstin Kennedy, Mariam Owen Ross, Clare Phillips, Sonnet Stanfill, Eric Turner, Florence Tyler and Marta Weiss.

Designing this exhibition has been a miraculously smooth creative process, in no small part to the excellent design team at our disposal, including Hikaru Nissanke, Jon Lopez, and Emily Priest at OMMX, Sebastian White and Tom Joyes at Kellenberger-White, Sam Britton at Coda to Coda, and Mike Breen at Media Powerhouse.

We are particularly thrilled about the bespoke films shot around the world that we were able to produce for the exhibition. The can-do spirit of Alice Doušová, Tetsuo Mukai, and Joe Almond at Zuketa were key to making this happen. A special thank you to Luke Halls Studio for its work on the final film. Jane Audas has been indispensable in coordinating this work.

The entire Exhibitions team at the V&A cannot be thanked enough for its professionalism and rigour in pulling off this project. A big thank you to Linda Lloyd-Jones who initially championed the project, and Daniel Slater for carrying it through to the opening. The core project team of Samantha King, Catherine Sargent, and Imogen Lyons has been the secret behind this exhibition's success. Clair Battisson, Eoin Kelly, and our entire Conservation Department, as well as Allen Irvine and our Technical Services team have been incredibly helpful working tirelessly to present the objects at their best.

Thank you to Asha McLoughlin and Bryony Shepherd for their guidance in the interpretation of the exhibition, and their diligence in ensuring the exhibition communicated its messages clearly to the broadest possible audience.

We have been fortunate to have had incredibly bright minds volunteer their time in the research phase of our exhibition, and we would like to thank Mina Song, Maude Willaerts, Ekta Raheja, Shruti Chhabra, Jekaterina Potasova for their early research work.

This beautiful book you are holding is due in no small part to the V&A publishing team's efforts: Kirstin Beattie, Coralie Hepburn, Tom Windross, Stella Giatrakou and Emma Woodiwiss, as well as the book designer Jonathan Abbott, copy-editor Mandy Greenfield and proof-reader Linda Schofield. Thank you to Jo Norman, who aside from her long-running support for the exhibition, was also our reader and provided valuable feedback on the content of the book.

This exhibition evolved from an initial concept which Helen Evenden and Norman Foster discussed with the previous Director of the V&A, Martin Roth, on the topic of the automobile. We remain indebted to their early thinking, passion for the subject, and the momentum they brought for a show of this kind at the V&A. Martin Roth's passing two years ago was a great tragedy, and we can only hope that we have done some justice to his initial vision. It is to our current director Tristram Hunt's great credit that he saw the potential in this exhibition and persevered with its production.

Last but not least, Esme Hawes, our research assistant – whose work far outstrips her job title – has been an indispensable force in the production of this exhibition. Her intelligence, creativity, and positive spirit have made putting this exhibition together a joy.

Johanna Agerman Ross is the Curator of Twentieth Century and Contemporary Furniture and Product Design in the Design, Architecture and Digital and Furniture, Textiles and Fashion Departments at the V&A. She founded the quarterly design journal *Disegno* in 2011, which she still directs, and was previously an editor at the monthly design and architecture magazine *Icon*.

Allison Arieff is the editorial director for SPUR (the San Francisco Bay Area Planning and Urban Research Association), an urban planning and policy think tank. She is also a regular contributor to the *New York Times*, and writes about architecture, design and technology. She is former editor-in-chief (and was founding senior editor) of *Dwell* magazine.

Lizzie Bisley is co-curator of *Cars: Accelerating the Modern World*. Previously a curator in the Furniture, Textiles and Fashion Department, she has worked on a number of exhibition and gallery projects at the V&A. These include the 2017 exhibition *Plywood: Material of the Modern World*, the Dr Susan Weber Furniture Gallery and the Europe 1600–1815 Galleries.

Laurence Blair is a freelance journalist who has reported from across Latin America for the BBC, *The Economist*, the *Financial Times* and the *Guardian*, focusing on politics, history, development and human rights. His book on South American history will be published by Bodley Head in 2020.

Brendan Cormier is a Senior Design Curator at the V&A and co-curator of *Cars: Accelerating the Modern World*. From 2014 to 2017 he was the lead curator of the V&A Gallery at Design Society in Shenzhen, China. In 2016 he curated the first Pavilion of Applied Arts at the Venice Biennale, with an exhibition called *A World of Fragile Parts*. Prior to working at the V&A, he was the managing editor of *Volume* magazine.

Esme Hawes is an assistant curator at the V&A and of *Cars: Accelerating the Modern World*. From 2017 to 2018 she was the assistant curator for the V&A Gallery at Design Society in Shenzhen, China, as well as *Robin Hood Gardens: A Ruin in Reverse* at the Pavilion of Applied Arts at the Venice Biennale.

Nanjala Nyabola is a writer and political analyst based in Nairobi, Kenya. She is the author of *Digital Democracy, Analogue Politics: How the Internet Era is Transforming Politics in Kenya*. She is also a frequent contributor to the BBC, the *Guardian*, *Foreign Policy*, *Foreign Affairs*, IRIN, *New African* magazine, Al Jazeera and others.

Oli Stratford is the editor-in-chief of the quarterly design journal *Disegno* and is also the head of editorial at *Tack Press*.

Georgina Voss is an artist, writer, and researcher. Her work focuses on the presence and politics of large-scale technology, heavy industry and complex systems. She is a Reader in the Design School at the London College of Communication, UAL, where she leads Supra Systems Studio; and is co-founder and director of consultancy Strange Telemetry.